African Women Trailblazers in STEM

Despite recent advancements, gender disparities in STEM disciplines remain a major concern worldwide, particularly in Africa. This book focuses on empowering women in STEM and highlighting how they have overcome the barriers they face in accessing and excelling in these areas, as well as proposing concrete solutions to bridge this gap. Africa possesses a vast pool of untapped STEM talent, especially among women: this book aims to support women who are thriving in STEM careers and who have provided opportunities and support for others, leading to a more diverse and dynamic workforce.

African Women Trailblazers in STEM: Pioneering Technological and Economic Advancement offers firsthand accounts of successful women in STEM fields across Africa, providing access to valuable data on economic innovation in the region. It sheds light on the challenges African women face in STEM and offers strategies to overcome them, showcasing the achievements of women in STEM and their contributions to the field. The book also explores the correlation between women in STEM in Africa and the Sustainable Development Goals, presenting case studies and data on the gender gap in STEM fields. Acknowledging the unique challenges women of diverse backgrounds face, including race, ethnicity, social status, and location, this book provides readers with a deeper understanding of the importance of inclusive empowerment strategies and how intersecting identities impact women's experiences in STEM fields.

This book can educate stakeholders and policymakers on the importance of funding programs that promote gender equality and women's participation in STEM fields. By offering evidence-based recommendations, the book can influence policy decisions aimed at creating more inclusive and supportive environments for women in STEM education and professions.

Women in Industry 4.0: Shaping World Foundations in Engineering, Business, and Sustainability

Series Editors: Tilottama Singh, Richa Goel, and Kyla L. Tennin

According to a World Economic Forum report, only 22% of AI professionals globally are female. This series will bring stories straight from the heart of these innovative women who have established themselves, to inspire millions and create a wave of impact. This will be a series of books and authors that will focus on sustainability through technology and entrepreneurship in Industry 4.0. The series will act as a global platform for bringing together organizations and women sharing their life cases, research, and grounded theory which will be an interesting read for engineers, businesses, government, researchers, and students working on startups and contributing to gender diversity.

Each book in the series is dedicated to illuminating women's research and achievements in key, targeted areas of the contemporary tech world and raising awareness of the pivotal work women are undertaking in areas of importance to our global community toward sustainable development. It will be a step in bringing forth the reality of what overshadows women in workplaces and the strategic attempts of organizations to pave the path for women's representation in companies and fields of engineering.

Books that focus on women in engineering and technology exploring and describing how women have greatly contributed to the world at large through these fields and their many subsections is a focal point of this series.

If you are interested in writing or editing a book for the series or would like more information, please contact Cindy Carelli, cindy.carelli@taylorandfrancis.com.

Women-Led Innovation and Entrepreneurship

Case Studies of Sustainable Ventures and Practices

Edited by Esra Sipahi Döngül, Serfie Uguz Arsu, Richa Goel, and Tilottama Singh

African Women Trailblazers in STEM

Pioneering Technological and Economic Advancement

Edited by Uzo Mkparu, Richa Goel, Tilottama Singh, and Jade Mkparu

African Women Trailblazers in STEM

Pioneering Technological and Economic Advancement

Edited by
Uzo Mkparu, Richa Goel,
Tilottama Singh, and Jade Mkparu

CRC Press
Taylor & Francis Group
Boca Raton London New York

CRC Press is an imprint of the
Taylor & Francis Group, an **informa** business

Designed cover image: Shutterstock

First edition published 2026
by CRC Press
2385 NW Executive Center Drive, Suite 320, Boca Raton FL 33431

and by CRC Press
4 Park Square, Milton Park, Abingdon, Oxon, OX14 4RN

CRC Press is an imprint of Taylor & Francis Group, LLC

Library of Congress Cataloging-in-Publication Data
[Insert LoC Data here when available]

ISBN: 978-1-032-88861-3 (hbk)
ISBN: 978-1-032-88974-0 (pbk)
ISBN: 978-1-003-54061-8 (ebk)

DOI: 10.1201/9781003540618

Typeset in Times
by Deanta Global Publishing Services, Chennai, India

Contents

Preface

"Why Africa?" "Why STEM?" "Why now?"

Because new technologies are reshaping our world at the speed of lightning-fast iteration, and African innovators are definitely part of the fray.

To say, in the 21st century, that technology has changed how we live would be something of an understatement. Innovations and improvements in science, medicine, and other STEM fields have not just woven their way into our daily patterns, but they have actively changed the patterns themselves. In fact, to describe modern life is to first mention the digital and technological revolutions that have happened in the past 20 years.

This book does not aim to cover old ground.

Instead, we want to shift focus and tell the stories of how STEM fields and innovations are actively beingshaped, and by whom, not just cataloguing the waves of change in a passive manner. It's clear that in every sphere of business, the proliferation of transformative technologies is not the news story: it's who's doing the change-making, who's doing the trailblazing to stay ahead of those changes and to make them happen.

Novelty and scale are no longer competitive advantages in an oversaturated space. With the sheer volume and speed of development, the positioning and visibility of your projects and achievements become crucial. It's these differentiating factors, alongside the actual innovation, that help your work to stand out from the noise. Taking charge of your own narrative is the best way to do this.

But with everyone posting on their own personal pages and websites, drawing attention to one's achievements and projects can be a fractured and frustrating process. This is why a book like 'African Women Trailblazers in STEM: Pioneering Tech' is important: we wanted to create a forum where these success stories can be catalogued in a centralised, easy-to-access way. This book meets a growing need for a platform where achievements and breakthroughs on the continent can be celebrated and rightfully positioned as leading developments in their fields. This positioning goes a long way towards strengthening your professional or company brand. We believe it is important to increase the visibility of practitioners who come from backgrounds that are underrepresented in mainstream conversations about innovations in STEM fields.

We hope that you enjoy the stories and case studies of the contributors featured herein and that their words broaden your perspective on how STEM can impact economic development and technological innovation across the continent.

Foreword

The Imperative of Women in STEM—A Personal and Continental Perspective

As a woman who has spent over two decades in STEM, leading multinational technology companies across the African continent, I have witnessed firsthand the incredible contributions African women have made—and will continue to make—in science, technology, engineering, and mathematics (STEM). From grassroots innovation to executive leadership, African women are pivotal to the future of our continent and the global STEM landscape. Yet, this is just the beginning.

Africa's future is undeniably bright. With over 40% of its 1.6 billion inhabitants under the age of 25, the continent boasts one of the youngest populations in the world. The rapid digitization of the continent positions it as a key player in the global economic and technological transformation. To fully unlock Africa's potential, however, we must ensure that **no gender is left behind**, especially in the critical fields of STEM.

THE IMPACT OF WOMEN IN STEM

The involvement of women in STEM is more than just a matter of equity—it's a powerful catalyst for innovation, economic growth, and social progress. According to the *World Economic Forum*, increasing women's participation in STEM fields could add $12 trillion to global GDP by 2025. Yet, despite the vast potential, women remain underrepresented in these fields globally, particularly in leadership roles. In Africa, women occupy only 30% of roles in science and technology, and the numbers drop further in executive positions. This disparity not only limits women's individual potential but also stifles broader societal progress.

Increased female participation in STEM has the power to drive innovation, boost economic competitiveness, and promote social equity. This book goes beyond merely outlining the challenges faced by women in STEM—it offers real solutions. It also emphasizes that **women's involvement in STEM is not just a matter of fairness—it is an imperative** for Africa's success.

A CALL TO ACTION: FOR GOVERNMENTS, BUSINESSES, AND CIVIL SOCIETIES

This book is not only for individuals—girls, women, or those already in STEM careers. It is a call to action for businesses, governments, and civil societies to recognize the broader societal and economic benefits of gender inclusion in STEM. It urges decision-makers to rethink policies, invest in education and mentorship, and create environments where women can thrive and lead.

The case studies in this book highlight the transformative impact of supportive policies and systems in enabling women to excel in STEM. From targeted scholarships and mentorship programs to inclusive workplace cultures, these solutions provide a roadmap for governments and businesses to follow. The future of Africa's technological and economic growth hinges on a commitment to nurturing talent in all its forms—especially by ensuring that **no woman is left behind** in the STEM revolution.

CONCLUSION

This book matters because it connects the dots between individual success stories and the broader societal impact of STEM education and inclusion. It's about changing the narrative and providing a platform for women to share their journeys, challenges, and triumphs. It's about creating a world where the leadership, skills, and insights of women in STEM are valued and leveraged to solve the world's most pressing problems.

Through their dedication and passion, the editors have curated a collection of powerful stories from women across Africa: women who are breaking barriers, innovating, and leading with courage and conviction. These individuals are not just participants in STEM; they are trailblazers, mentors, and role models. Their stories illuminate the importance of representation and the profound impact that women have on shaping industries, institutions, and communities.

The path forward is clear: if we want to ensure a prosperous, innovative, and sustainable future, we must embrace the power of gender-diverse STEM teams. Africa's women are at the forefront of this change, and this book invites you to join the conversation, contribute to the solutions, and be part of the movement that will shape tomorrow.

Jane Egerton-Idehen
MD/CEO
Nigerian Communications Satellite Ltd (NIGCOMSAT)

About the Editors

Uzo Mkparu is the Founder and Chief Executive Officer of Techscale Consulting Ltd UK, a digital innovation and AI consultancy, and Techscale Technologies Ltd. An award-winning Chief Executive with over 25 years' experience, spanning financial services and other industries, she specialises in strategy, business transformation, and innovation-led growth with a focus on emerging technologies. She has excelled in leadership positions and achieved notable success working for top-tier companies such as Oracle and ICL Group in the United Kingdom and leading banks in Nigeria. She holds an MBA from Cranfield University, UK, and a Master's degree in Information Systems and Technology from City University, London, and has acquired numerous certifications in business and technology.

Uzo is an AI Governance Fellow and is passionate about leveraging technology and positioning Africa to harness the economic potential of emerging technologies and data. She serves on several boards including the Advisory Board of Data Science Nigeria and Preston Associates International Development. She is an accomplished Mergers Integration Director, Consultant, Trainer, Global Conference Speaker and also spent multiple years as a regular guest business speaker at Pan-Atlantic University. Uzo has several publications to her name, focusing mainly on emerging technology. She was recently recognised as one of the 100 Leading Women in Nigeria. She is an Honorary Senior Member (HCIB) of the Chartered Institute of Bankers of Nigeria and a member of Boardroom Africa.

Tilottama Singh is a certified HR Analyst from IIM and a proficient Academic, Researcher, and Trainer with more than 14 years of experience in the field of Human Resources and Work Dynamics. Presently, she is working as Head of Department in the School of Management, Uttaranchal Institute of Management, Uttaranchal University, Dehradun. Her areas of expertise are Human Resources and Strategy, and her teaching concentration includes human resources, strategy, and law. She started her career in the hotel industry with Leela Kempinski in the corporate head office in Human Resources and Training. Later, she joined the education sector with topmost universities IMS Unison Group and Amity University, the leading university in Pan Asia. She has obtained her Master's in Human Resources Management with Distinction. Prior to that, she graduated in economics with honours and law with distinction, with an added vocational course in mass communication and analytic certification from IIM. Having an enriched research portfolio with Scopus and international refereed journals, book chapters, and conference presentations, she has been awarded the Best Presentation Award in international conferences like IEOM, MIDAS, and many more. She also serves as a member of AIMA and acts to lead and liaise between the student community and the industry delegates, with a keen interest in training and building sustainable models for business and society.

Richa Goel is an Associate Professor of Economics and International Business at SCMS Noida, Symbiosis International Deemed University, Pune. She is a gold medallist in her master of economics with dual specialisation at the master level accompanied by an MBA in HR and also with dual specialisation at the graduate level in economics and law. She has a PhD in Management, where she has worked for almost six years in the area of diversity management. She has almost 20+ years of experience in academics along with authoring more than 50+ research papers for publication in reputed national and international journals, in addition to hundreds of papers for international/national conferences and symposiums. She has written more than 25 book chapters for leading publishers, and she has served as a member of the review committee for conferences and journals. She is also currently a Series Editor for two different series for Taylor and Francis.

Jade Mkparu is a management consulting analyst and an award-winning author and essayist. She is also the Founder and CEO of BRAIDAR: a lifestyle concierge service for young black girls studying in the UK, connecting them to nearby braiders and to a sense of community. Jade was also celebrated by Powerful Media's Future Leaders Magazine 2024–2025 as one of the 150 Top Black Students in the UK. She was a volunteer with UN Women UK during the Commission on the Status of Women 68 and has a joint English and Management undergraduate degree from the University of St Andrews. Jade began writing as a biweekly film journalist in 2011, as the youngest writer on the FilmGuide and FilmXtra teams for Genesis Cinemas and Filmhouse Cinemas in Nigeria. She has also been published in the YoungWriters 2019 Poetry Escape anthology and the Hurtwood Photoverse publication, and was a winner of the latter's eponymous competition. She also won First Prize in the Chairman's Award Essay competition for two consecutive years at 12 and 13 years old, first in her year group category and then the whole-school prize of $2,000. She has been a panelist speaker at student entrepreneurship events, has given keynote graduation speeches at her alma mater in Lagos, and is a mentor to young girls who want to further their writing careers. Jade is also the Co-Founder of Lagos Family Fun, an online tourism and lifestyle brand for families in Lagos. She is also the creator and host of the live radio show 'BRAIDAR: The Business of Beauty', in which she analyses the innovation strategies and positioning of beauty and fashion brands, as well as emerging Gen Z consumer trends that they could leverage. She has previously completed a research project on the potential of blockchain for collaborative storytelling and has internship experience with Accenture in management consulting and with Founder's Intelligence in innovation strategy consulting. She was also part of the Penguin Random House UK and Creative Access mentorship programme, where she was mentored by the Director of Strategy.

Contributors

Rolayo Akhigbe
Senior Executive, Financial Services, Mentor, Angel Investor
Lagos, Nigeria

Khawaja Sazzad Ali
Research Associate
Research and Policy Integration for Development (RAPID)
Department of Economics, North South University
Dhaka, Bangladesh

Stella Ihenacho
Co-Founder and Agile Faculty Lead
NSTAR Technologies
Atlanta, GA

Adaeze Iloeje-Udeogalanya
Founder and CEO
African Women in STEM
Princeton, NJ

Sandra Ilogu
Third Party Risk and Digital Capabilities
Consultant Neurodiversity and Health Equity Advocate
NJ

Oreoluwa Lesi
Founder and Executive Director
Women's Technology Empowerment Centre—W.TEC
Lagos, Nigeria

Ngozi Megwa
Executive Leader in Digital Payments and Commerce
Global Financial Services
Dubai United Arab Emirates

Pretty Mubaiwa, PhD
Gender Intentionality and Inclusion Consultant
South Africa/Zimbabwe

Dativah Bideri Mukamusonera
Founder and Executive Director
Girls Tech Ventures
Kayonza, Rwanda

Chiso Ndukwe-Okafor
Executive Director
Consumer Advocacy and Empowerment Foundation
Lagos, Nigeria

Eleanor Nwadinobi
International President
Medical Women's International Association
Geneva, Switzerland

Eneni Oduwole
Founder/CEO
Alter-Ed Limited
Lagos, Nigeria

Bimpé Olufemi
Digital Strategy Director
Junior Achievement Africa
Ghana

Quazi Tafsirul Islam
Research Associate
Research and Policy Integration for Development (RAPID)
Department of Economics
North South University
Dhaka, Bangladesh

Md. Shamim Talukder
Research Associate
Research and Policy Integration for Development (RAPID)
Department of Economics
North South University
Dhaka, Bangladesh

1 Cultural Shifts

The New Face of African Women in STEM in Africa and the Diaspora

Adaeze Iloeje-Udeogalanya

"Cultural Shifts: The New Face of African Women in STEM in Africa and the Diaspora" discusses the roles, achievements, and contributions of African women in STEM, how these are evolving, and what is needed to support and accelerate this evolution. This chapter highlights a woman's personal journey from a young girl considering STEM to the Founder of African Women in STEM, along with real-life experiences of the women in her community, to provide a comprehensive view of the ongoing challenges and opportunities for impactful change. This chapter is written for African women in STEM, their peers, allies, and advocates who want to collaborate and amplify their collective impact to create transformations that will ripple through future generations.

The world is not static, nor are African women in STEM, and consequently neither should culture be.

INTRODUCTION

A quiet cultural revolution is taking place.

African women, in the continent and beyond, are challenging stereotypes, reshaping narratives, and redefining their roles in traditionally male-dominated fields and male-centered societies. The heart of the revolution lies in the realization that, contrary to the existing narratives, they have the power to transform their careers and lives and impact their communities and the world.

I have always been passionate about STEM. I loved my engineering career and was fueled by the desire to create innovative solutions that could change lives. However, at various stages of my career and in different workplaces, I encountered obstacles that impeded my progress. These hurdles were deeply rooted in societal and cultural constructs that define a woman's supposed role and what she could aspire to become.

For decades, I was repeatedly told, both overtly and subtly, that my place was not in STEM.

The bar was set impossibly high for me in my career, and I was required to prove my worth over and over again. The feedback I received was "your prior achievements

DOI: 10.1201/9781003540618-1

may have been a fluke" or "show us, again, that you can do it." I was expected to fit into predefined roles and be less ambitious. I was pressured to temper my aspirations and limit myself in order to fit societal and cultural constructs predetermined for a woman, an African woman.

The cultural expectations weighed heavily on me to the point that I nearly walked away from a career I love. I did not surrender, however, and resolved to stay in STEM and work to dismantle the very biases and constructs that sought to push me out. What began as a personal determination to defy cultural norms turned into a mission to help other women facing the same struggles.

This chapter explores my journey from a deeply curious young girl to the Founder of African Women in STEM, along with real-life experiences of the women in my community and women in STEM at large, to provide a comprehensive view of the ongoing challenges as well as opportunities for impactful change.

THE BEGINNING

"I want to be an astronaut," I blurted out one Saturday morning as I moved to sit beside my father on the couch.

"Why do you want to be an astronaut?" my father asked, putting down his book.

What does he mean by "why?" my six-year-old self wondered. Does he not know what an astronaut is? Are adults not supposed to know everything?

"Well, I want to be the first girl on the moon."

"Great!" my father responded, and I beamed.

"That's wonderful. Now, what do you need to do to become an astronaut?"

Another question? All I wanted was applause, a spacesuit, and a bunch of the gadgets astronauts have on TV.

"Mmmm, I don't know," I shrugged.

"Well then, your homework is to go and find out what it takes to be an astronaut."

I did not quite like where this was going, but little did I know that this was only the first of what would feel like a million quests my father would send me on.

I turned my attention back to the television to watch more sci-fi cartoons, determined to figure out what I needed to do to become an astronaut.

I did not know it then, but the fact that my father supported my interest, engaged me in conversation, and challenged me to explore things further was so far from the norm, it was a *unicorn experience*!

For many young girls interested in STEM and women in STEM in Africa, such support is rare.

Imagine trying to climb a mountain with a backpack full of rocks. That is often what it feels like for these women and girls. Their journey is a constant battle to overcome challenges that may be invisible to most.

And what are these "rocks" they carry?

- Pervasive gender stereotypes that suggest women and girls do not belong in STEM fields.
- Unconscious bias that pushes the notion they are less than and undermines their contributions.

- School and workplace discrimination, ranging from subtle microaggressions to outright hostility, that turn their environments into minefields.
- Lack of mentorship leaves them without role models, making it harder to envision themselves entering or thriving in STEM (you can't be what you can't see).
- Opportunity and wage gaps are real and persistent and are a consequence of the inadequate recognition of their achievements.
- When you constantly hear "you can't," it is difficult not to internalize that message and form self-limiting beliefs.

NAVIGATING THE STEM LANDSCAPE: CHALLENGES

Gender Stereotypes

In Africa and the African diaspora, gender stereotyping is a significant barrier to the involvement of young girls in STEM. Strongly held beliefs and narratives suggest that careers in STEM fields do not align with the primary or traditional gender roles of caregiver and homemaker. These beliefs are often perpetuated through societal norms, media portrayals, the education system, and within families.

Many cultures have deeply ingrained gender roles that reinforce harmful misconceptions, such as "girls are not good at math" or "engineering is better suited for men." Girls and young women are discouraged from being involved in science and technology programs or exploring STEM disciplines. It is considered inappropriate or unnecessary for them to engage in "hard sciences" or technical fields, reinforcing the notion that their intellectual contributions in STEM are less valued.

Studies have shown that these narratives begin to impact young girls as early as age six.[1,2] According to the 2019 UNESCO report on women in STEM in sub-Saharan Africa, fewer than 20% of girls expressed interest in pursuing careers in engineering or technology by age 12, with many citing societal expectations as a deterrent.[3] Jennifer Nansubuga Makumbi puts it best in her book *A Girl is a Body of Water*, where she expresses that generally, African folktales about women portray them as either dangerous, manipulative, vain, or submissive, rather than complex, strong, powerful, brilliant, or autonomous individuals.[4]

For generations, our society has been plagued by these harmful gender norms, which, interestingly, may be masking a deeply rooted fear of what women might accomplish if they gain confidence and the opportunity to express it. As a consequence, keeping women submissive through gender stereotyping has become a tool to maintain traditional gender roles. Any thoughts, ideas, or expressions of their talents to build, create, innovate, or heal are, at best, discouraged, and at worst, punished or completely extinguished.

Unconscious Bias

My career aspirations were an ongoing discussion throughout my childhood. I went from wanting to be an astronaut to wanting to be a botanist (*though I changed my mind quickly because the name sounded boring*), and by the time I was 13 years old, I zeroed in on engineering, pharmacy (*medicine was out since I could not stand the sight of blood*), and economics (*because I thought it would be interesting to study human behavior from this perspective*).

I am grateful for those discussions because they helped guide my decisions and examine the reasoning behind them. One such conversation was crucial in shaping the person that I am today.

As a kid, I longed to leave home, explore the world, and, most importantly, have the freedom to do whatever I wanted!

Both my parents were in academia at one of the most prestigious universities in Nigeria, the University of Nigeria Nsukka (UNN), so it made perfect sense that they would want all their children to attend college there. However, I had other ideas. I wanted to be as far away as possible from my parents' circle of influence. So I researched engineering programs that were not offered at UNN and settled on computer engineering.

A few months after I finished secondary school, I was sitting at the table filling out my college application forms, the Joint Admission and Matriculation Board application forms, famously known as JAMB in Nigeria. Ordinarily, my father would have been involved, but thankfully he was out of the country and could not interfere with my plans. Unfortunately, I did not factor in the other key person, my mother, and for some reason, she picked that exact moment to walk into the dining room.

"Have you settled on your course choices?" she inquired.

"Yes," I said and listed out computer engineering and two other choices.

My mom paused, then asked, "Why, computer engineering?"

Since I could not tell her my real reasons, I began to share what I had read in recent articles about computers being the driving force of the future and explained how I wanted to be at the forefront and leading this revolution.

"What do you want to do when you graduate with this degree?" she probed.

"I want to enhance the lives of women in rural areas by providing them with accessible and affordable energy sources for cooking and other household energy needs. I want to reduce pollution and improve their overall health."

She regarded me silently for a moment, then said, "That sounds exactly like what your father is doing at the Renewable Energy Center." My father is a mechanical engineer.

And just like that, my well-crafted plan unraveled.

My plan to "run away from home" fell apart, but my unwavering focus on pursuing an engineering degree and changing lives through my work was not impacted because I had incredible role models who helped me develop a strong foundation and a strong "Why."

This last part is important because I experienced a lot of ridicule from friends, family members, and perfect strangers, mostly male, whenever I answered the question "What are you going to study in the university?" with "Mechanical Engineering."

People would laugh and ask me if I realized I was female or make statements such as "women do not study mechanical engineering" or "the program is one of the toughest to get into and you do not stand a chance."

Rather than discourage me, these remarks increased my resolve to prove the doubters wrong and show them that I could do this seemingly impossible thing, and I did! I got into the program on merit and graduated in the top 3% of my class!

Sadly, not everyone is as fortunate as I was.

A year or two before I was admitted into the university, my neighbor's cousin moved in with them to prepare for and take the university entrance exam. This was a common practice, as the exam preparation programs on or close to the campus were top-notch. She was able to apply for and get into the mechanical engineering program. Unfortunately for her, her father found out about it. He was so outraged that she had done this without his permission that he threatened to cut her off. She had no choice but to turn down the admission, retake the entrance exam, and reapply to a more "appropriate" course for women. She lost time, a whole year, and most likely a lot more—her dreams and ambitions. I do not know what she ended up studying, but I can imagine she had lots of regrets about it.

Well-meaning people derail the dreams and aspirations of girls and young women who are interested in entering STEM fields, all the while believing that they are being helpful. What they fail to realize is that their actions, though couched in a desire to lighten the girl's load as she climbs the mountain, actually serve to, in fact, add boulders into her bag, making it impossible for her to ever reach the top of that mountain.

Cultural norms often shape unconscious bias.

In most societies, men typically hold leadership roles. Male employees are favored for high-profile projects and roles because the assumption that female employees are less capable or will not be as committed as men is pervasive. These biases are amplified by cultural norms that expect women to prioritize home life, resulting in fewer opportunities for professional growth.

Studies show that both teachers and parents are more inclined to encourage boys rather than girls to engage in STEM activities, often based on the assumption that boys naturally possess a stronger aptitude for these subjects. This unconscious bias sends a message to young girls that they do not belong in STEM and further contributes to their lack of interest in these fields.[5] It can also, as was the case with my neighbor's cousin, force them to abandon their potential to innovate and make a difference in the world.

Workplace Discrimination

I worked in predominantly male organizations where the number of female engineers was so few I could count them on one hand, with fingers to spare. Sometime in the middle of my career, I found myself at yet another such organization. I had an experience that I initially assumed was a positive and welcome change but soon realized it was not.

My first year there, I was invited to take part in activities celebrating International Women's Day. I had not heard of this event before, but I was excited to participate and connect with other women in the organization.

During the first part of the day, senior leadership (who were all male) participated. Later, there was a women-only forum with a featured guest speaker. This provided the women with a chance to speak candidly about the challenges they experience in the workplace.

A young woman, just two years out of college and recently married, shared an issue that had been weighing heavily on her mind. She spoke of how the more senior women on her team were badly treated, and it raised deep concerns for her future. In particular, she recounted the experience of a senior-level employee who, after returning from maternity leave, faced dramatic shifts in how the predominantly male team treated her. The woman's ideas and contributions were consistently dismissed by the team. She was excluded from key projects. What bothered this young woman was that the treatment seemed directly linked to her colleague's decision to have a child.

As she shared her thoughts, it became evident just how much the situation had affected her. She admitted it made her feel anxious about what she might face if she chose to start a family of her own. The possibility of facing discrimination, exclusion, or unfair treatment simply for choosing to have children made her question whether starting a family was even worth it.

Hearing her story was heartbreaking. It is tragic to think that workplace discrimination could be so profoundly detrimental as to cause women to choose between work and starting a family. It is neither right nor fair for the workplace to have such an impact on a woman's life choices. For women, like their male counterparts, career success and starting a family should not be mutually exclusive. Yet, given what this young woman had experienced firsthand, her fears were entirely justified.

African women in STEM across the diaspora, particularly in the United States, face distinct and multifaceted workplace discrimination. Working in male-dominated industries is challenging enough, but they must also navigate additional cultural and racial barriers that routinely impede their career advancement and professional success. These intersecting challenges often lead to their potential being capped. A confluence of gender, race, and cultural factors conspires to keep them undervalued, overlooked, underpaid, and typically excluded from positions of power. Despite making significant contributions, they are frequently required to prove their worth in spaces not designed for them to excel.[6,7]

Invisibility Barrier

While invisibility may be a superpower in sci-fi comics and movies, sadly it is not for African women in the workplace. One of the most pervasive forms of discrimination African women face is being rendered invisible in the workplace. Many women report that, in spite of their qualifications, they are more likely to be passed over for promotions.[8] They have their ideas credited to others, often without them being acknowledged. It is deeply frustrating for these women to feel unseen, especially knowing they often put in considerably more effort than their peers.[9]

In many instances, their contributions are not viewed through the same lens as those of their male colleagues. They encounter assumptions that they are less competent, less intelligent, or less capable due to preconceived notions about their race, cultural background, gender, or education.[10] Their work is frequently subjected to harsher criticism, which creates an environment where their achievements are not acknowledged, regardless of how significant their contributions are.[11] Consistently being overlooked can lead to stress and affect their overall sense of satisfaction in both their careers and personal lives.

A couple of years ago, I was hired by an organization specifically for my extensive skills and experience in the industry, as well as the scope and scale of the projects I had successfully executed. However, throughout my time at this organization, the majority of the team members across ranks, from senior executives to the newest male hire, were collectively working to undermine my contributions.

One morning, while I was preparing a risk management plan, my phone rang.

It was Gray.

He recently transitioned from another division within the company. He needed information about a project he had been assigned to. Just before he ended the call, he paused and asked,

"By the way, what do you do at the firm?"

I told him what I did, what my expertise was, and the experience I had prior to joining the firm.

He seemed surprised and admitted he had assumed I was an admin.

That hurt.

I ended the call and sat there in shock for what seemed like an eternity.

I reflected on the fact that I executed critical work successfully and delivered value and results, and yet I had not received credit. I reflected on the fact that while I was on maternity leave, every single member of my project team had been promoted and given a raise for the project's success. Everyone except me.

My achievements were deliberately hidden, and it was clear they had succeeded because I had slowly become invisible in the group.

My experience is not unique to me. There are grave consequences to making someone feel unseen by disregarding their contributions and expertise. Actions like these can damage a person's self-worth and affect their health and well-being.[12]

The Hidden Cost Of Bias

I moved to the United States to obtain a master's degree and pursue a career in the oil and gas pipeline industry. I was intrigued by how advanced the industry was and the opportunity to design and build facilities I had only conceptualized. I could not wait to start creating and innovating till date.

The misconception I had was that development meant equal opportunity. Though I excelled in many ways, showcased my unique value, and contributed to higher profitability in multiple organizations, I was still impacted by some of the same issues faced by women in STEM in Africa and much more.

Company culture and societal expectations in the United States cause African women in STEM to contend with racial and cultural discrimination, motherhood bias, being undervalued, and lacking representation. The impact of bias goes beyond missed opportunities at work. It eats away at their confidence and threatens their feelings of safety. Always having to prove that who they are or that their life choices do not make them any less capable or committed is both unfair and exhausting. The danger is that these experiences have become normalized.

Research shows that bias and discrimination have a significantly negative impact on black women, especially in STEM fields.[13] The "weathering" effect, a term created by researcher Arline Geronimus, explains how long-term stress from racism can make black women age faster and have more health problems.[14] This constant exposure to unfair treatment affects how they handle emotions and leads to health conditions like anxiety, depression, high blood pressure, and even higher chances of developing certain cancers.[15,16] Workplace bias often forces black women to deal with these stresses alone, without enough support from their organizations, which makes the health risks even worse.[17]

Additional consequences to consider are that clinging to outdated practices wastes the potential of an entire segment of the workforce, causing organizations to stagnate, families to struggle, and communities to miss out on the brilliance of women. Ultimately, the world is robbed of talent and innovation that could transform the future.

Given what I have shared, it is therefore evident that supporting girls' and women's interest in STEM is crucial. Equally important is ensuring that when they enter into STEM careers, they not only succeed but also have the support to sustain long-term careers.

The time for change is not tomorrow. It is now.

SUCCESS STORIES: INSPIRING THE NEXT GENERATION

History, as well as current trends, has shown that African women in STEM across the world are visionaries, innovators, and trailblazers. Many women in Africa and the diaspora have broken through these prevalent barriers to achieve remarkable success in STEM fields that were once inaccessible to them. These trailblazing figures are leaders in their respective disciplines and serve as role models for future generations of women. Their stories reflect the power of resilience, innovation, and a shared commitment to not only excel in their fields but also uplift others. They have demonstrated that with perseverance, talent, and support, anything is possible.

The pioneers and emerging leaders featured here have achieved personal success and paved the way for countless others, inspiring the next wave of African women in STEM.

Pioneers

Wangari Maathai

Wangari, a Kenyan environmental and political activist, is celebrated for her pioneering contributions to science and environmental conservation. In 2004, she became the first African woman to receive the Nobel Peace Prize.[18] Her scientific accomplishments were groundbreaking; in 1971, she became the first woman in East and Central Africa to earn a doctorate, obtaining a PhD in veterinary anatomy from the University of Nairobi. She further broke more barriers by becoming the first female professor in Kenya. Beyond her academic achievements, Maathai established the Green Belt Movement in 1977, a grassroots organization responsible for planting over 30 million trees across Africa. Her efforts in sustainable development and environmental preservation made her not only a pioneering scientist but also a symbol of resilience and leadership in the fight for ecological and social justice. Maathai's work continues to inspire women and environmentalists around the world.

Francisca Nneka Okeke

Francisca is a Nigerian physicist celebrated for her pioneering research in atmospheric and space physics. She was the first woman to lead the Department of Physics and Astronomy at the University of Nigeria, Nsukka, and later served as the Dean of the Faculty of Physical Sciences from 2008 to 2010. Her research focused on the equatorial electrojet phenomenon in the upper atmosphere, a key area of study in understanding space weather and its effects on Earth. Okeke's contributions to science have been recognized globally, and she was honored with the L'Oréal-UNESCO For Women in Science Award in 2013. Her work not only advanced the field of atmospheric physics but also paved the way for other women in Nigeria and beyond to pursue careers in science. Okeke remains a role model for aspiring female scientists, particularly in fields where women are underrepresented.[19,20]

Winifred Byanyima

Winifred originally trained as an aeronautical engineer and made history by becoming the first Ugandan woman in this field. She studied aeronautical engineering at the University of Manchester and went on to earn a master's degree in mechanical engineering from Cranfield University. She was a flight engineer for Uganda Airlines. However, her career trajectory shifted as she became a prominent politician and human rights advocate. Byanyima's leadership extended beyond engineering and politics; she is currently the Executive Director of UNAIDS and an Under-Secretary-General of the United Nations, where she is deeply involved in global health issues, particularly the fight against HIV/AIDS. Her diverse career path showcases the breadth of her skills and her commitment to improving lives, not only through science but also through advocacy for social justice. Byanyima is an exemplary figure of how women can excel in multiple fields and make significant contributions to society.[21,22]

Emerging Leaders

Joy Adowaa Buolamwini

Joy, a Ghanaian-American computer scientist and digital activist, is renowned for her work in addressing racial and gender bias in artificial intelligence (AI). She founded the Algorithmic Justice League to raise awareness about the biases embedded in AI systems, particularly in facial recognition technologies. Her groundbreaking research revealed how these technologies often fail to accurately recognize people of color, especially women, sparking a global conversation about ethics in AI development. In

2019, Buolamwini was named to *TIME* Magazine's 100 Next list, recognizing her as a rising leader in technology. Through her work, she continues to challenge tech companies to create more equitable and unbiased AI systems, demonstrating the importance of diversity and representation in the field of computer science.[23] I had the honor of meeting her at the Grace Hopper Celebration 2024, hosted by Anitab.org.

Chinazom Arinze

Chinazom, also known as "AutoGirl," is a Nigerian advocate for increasing diversity in the automotive and tech industries. She founded the AutoGirl initiative in 2019 to empower women to pursue careers in the automotive sector, a field traditionally dominated by men. Arinze, a law graduate, turned her passion for cars into a thriving business that includes vehicle rentals and sales. She is committed to increasing female representation in STEM fields and actively organizes workshops and mentorship programs to encourage young women to explore opportunities in automotive technology. Through her work, she has inspired many women to break into industries where they have historically been underrepresented, fostering a new generation of female engineers and tech professionals in Nigeria.[24]

Angela Udongwo

Angela, a Nigerian-American researcher, has made notable contributions to the field of medical imaging. She is currently a third-year medical student at Temple University in Philadelphia, Pennsylvania, USA. Her groundbreaking research focuses on the impact of black hairstyles on medical imaging results, highlighting the importance of diversity in medical research and technology development. Udongwo's work addresses how certain hairstyles, common among people of African descent, can affect the accuracy of medical imaging technologies. Her research is vital in improving healthcare outcomes for individuals with diverse hair textures and styles, and it underscores the need for inclusive practices in medical research. Through her work, Udongwo is not only advancing science but also advocating for more equitable healthcare for underrepresented communities.[25]

These women, both pioneering figures and emerging leaders, represent the diverse and impactful contributions of African women in STEM. Their work spans a range of disciplines, including environmental science, physics, artificial intelligence, automotive, and medical research, demonstrating the transformative role that African women are playing in shaping the future of STEM.

When society ignores the immense contribution, aptitude, and ability of African women in STEM, the impact on the women is immense, but the loss that society experiences can be tremendous.

THE NEW FACE OF AFRICAN WOMEN IN STEM: REDEFINING THE NARRATIVE

The landscape is changing, and the numbers tell an exciting story.

In several Southern African countries (Botswana, Eswatini, Namibia, and South Africa), girls' secondary education enrollment is higher than that of boys, indicating a positive trend in early STEM education access.[26]

Forty seven percent of science, technology, engineering, and mathematics graduates at African universities, at both undergraduate and postgraduate levels, are women.[26] This represents the highest proportion of female STEM graduates globally. The high proportion of female STEM graduates in Africa (47%) compared to other regions suggests a positive trend. For comparison: Asia and South America each have 41%, Europe follows at 42%, North America at 39%, and Oceania at 38%.

Initiatives like the Economic Commission for Africa's (ECA) *Connected African Girls Coding Camp and Climate Change Adaptation Hackathon* involve young female engineers and innovators in creative projects leveraging emerging technologies to tackle climate change.[27] The Connected African Girls initiative, launched in 2020 by the Economic Commission for Africa (ECA), has trained approximately 25,000 girls in artificial intelligence, coding, gaming, and other digital technologies.

African women are increasingly taking on leadership roles in global tech initiatives. For example, women like Neema Iyer, founder of Pollicy (a feminist technology collective in Uganda), have been appointed to Meta's Global Women's Safety Advisory Board, contributing to shaping global tech policies.[28]

According to UNESCO, the percentage of female researchers in sub-Saharan Africa (31.8%) surpasses the global average (28.8%),[29] showcasing a commitment to promoting gender diversity in research and encouraging women to pursue careers in STEM, despite the challenges they face.

These statistics reflect the progress in educational opportunities for women and girls, as well as the resilience and ambition of African women in breaking barriers in male-dominated fields such as science and technology. African women in STEM are conducting groundbreaking research, developing innovative solutions to African and global challenges, and building powerful networks and communities. In recent years, female tech founders have been emerging across Africa's tech ecosystem, making significant strides and outperforming their male counterparts.[30,31]

The overall trend is positive, with more African women and girls entering STEM fields, taking on leadership roles, and driving innovation both in Africa and globally.

African women in STEM are not just participating, they are leading.

BUILDING NETWORKS AND COMMUNITIES

I have two decades of experience as an engineer and leader in the energy industry. Throughout my career, and especially as I became more senior, I found myself increasingly isolated. I did not have mentors, sponsors, and advocates who understood me and could provide a success roadmap that highlighted my unique abilities and talents, as well as my background and culture. I did not have a playbook for how to excel in the environments I found myself in, given everything I am—woman, black, African, immigrant, wife, and mother.

I was constantly expected to assimilate; to fit into a mold that was not made for me. The examples of who I was "supposed" to become were presented clearly, but they did not reflect me. I am not white, and I am not male. I do not play golf, drink beer while watching football on Sundays, or go hunting with the guys. Yet, success in my career—or even a shot at leadership—appeared to be available only to those who could conform to those very narrow standards. The culture was one where those who did not fit in were subtly and consistently excluded. Opportunities always seemed just out of reach for anyone who could not or would not mold themselves to fit.

The unspoken message was clear: if you are not part of the existing culture, success is not for you. I had to navigate the deep frustration and isolation that came with being visibly different in spaces where they chose not to recognize my potential and required me to play by rules that were not inclusive. The constant pressure to shrink who I was and become something I was not was exhausting. For women like me, the journey is less about growth and more about constantly fighting to survive in a system that was not designed for us to thrive.

I refused to let this be my story, but I know too many others have been crushed by the weight of these expectations. This culture, so out of touch with what true leadership and inclusion mean, has left so many of us questioning whether we would ever truly succeed.

The turning point, for me, came when a woman who was early in her career in my industry reached out to me for guidance, sharing struggles that mirrored my own experiences. Our conversation underscored the systemic nature of the issues many African women in STEM face in the workplace. I recognized that cultural expectations—whether societal, organizational, or systemic—were not just my own barrier but a collective one for African women in STEM. I was inspired to create an organization that offers a supportive community and a safe space for women to learn, thrive, and rise together. I set out on a mission to ensure no other woman feels the need to settle or give up her vision because of the challenges of navigating a career without the right support. So, in the fall of 2021, I founded my organization, African Women in STEM.

African Women in STEM is on a mission to empower women to excel in STEM and build a lasting legacy. We offer programs, resources, and a supportive community to help them take ownership of their careers, reach their full potential, and succeed.

The power of community and networking cannot be overstated when it comes to career advancement and accessing opportunities. In a world where 85% of jobs are filled through networking and 70% of jobs are never publicly advertised,[32] the value of who you know and who knows you is clearly evident. Networking provides an opportunity to make genuine connections and develop relationships that can lead to better career opportunities. Being part of a supportive community gives you a chance to tap into collective wisdom, shared experiences, and powerful connections that can accelerate your career.

However, despite the incredible benefits of networking, only 35% of women actively use networking to further their professions.[33] This gap represents missed opportunities, particularly in industries where women are already underrepresented.

Today, African Women in STEM has evolved into a dynamic organization that empowers women to excel, not just by providing support but by fostering cultural shifts. It is challenging the outdated societal expectations that have long defined women's roles and creating a new narrative—one where African women are recognized, valued, and celebrated for their contributions in STEM and to the global community, giving women a fighting chance.

EMBRACING THE FUTURE: YOUR ROLE IN THE REVOLUTION

The future for African women in STEM is bright, full of opportunities, and holds limitless potential. We can shape the global STEM ecosystem by leveraging our collective efforts to drive cultural shifts.

African women in STEM.

A new dawn has come.

Now is your moment.

Your contributions matter.

Embrace your power and take an active role in driving this transformation. The work we do today can set the stage for generations of trailblazers to come and build a world that champions innovation and excellence regardless of the source.

As we stand at this exciting crossroads, it is clear that a cultural revolution is underway. But this revolution needs you. Whether you are an African woman in STEM, a peer, an ally, a sponsor, or an advocate, your support is crucial. Engage in this mission and stand with us, so together we can create a future where African women lead, innovate, and thrive in STEM fields globally.

Here is how you can get involved (BASC©):

- **B**e a mentor, offer support, reach out, and lift others up.
- **A**dvocate for change and push for policies that support African women in STEM.
- **S**hare your story and inspire others. Do not underestimate the power of your voice.
- **C**hallenge stereotypes and speak up against bias and discrimination when you see them.

The world is not static, nor are African women in STEM, and consequently neither should culture be.

REFERENCES

1. How do gender stereotypes impact girls' interest in STEM subjects. (n.d.). Retrieved from https://www.ncbi.nlm.nih.gov/pmc/articles/PMC7104893/
2. Stem Women. (n.d.). New study suggests STEM gender stereotypes begin as young as age six. Retrieved from https://www.stemwomen.co.uk/
3. UNESCO. (n.d.). Unlocking Africa's potential by investing in STEM education. Retrieved from https://www.unesco.org/en/articles/unlocking-africas-potential-investing-stem-education
4. Makumbi, J. N. (n.d.). *A girl is a body of water.* Publisher.
5. What's the context? (n.d.). Retrieved from https://www.thegist.edu.au/educators/girls-in-stem/whats-the-context/
6. Women in the Workplace 2024. (n.d.). Retrieved from https://womenintheworkplace.com/
7. Harvard Business Review. (2018). How Black women describe navigating race and gender in the workplace. Retrieved from https://hbr.org/2018/03/how-black-women-describe-navigating-race-and-gender-in-the-workplace
8. Women are advancing in the workplace, but women of color still lag behind. (n.d.). Retrieved from https://www.brookings.edu/articles/women-are-advancing-in-the-workplace-but-women-of-color-still-lag-behind/

9. African diaspora women and African development. (n.d.). *In The Palgrave handbook of African women's studies.* Retrieved from https://link.springer.com/referenceworkentry/10.1007/978-3-319-77030-7_117-1
10. Iloeje-Udeogalanya, A. (2024, May 11). Cultural identity is key to unlock success in STEM [Video]. *TEDx.* Retrieved from https://youtu.be/xYcVJJKBQrY?si=iSGjCqpOtszXMgPN
11. Smith, A., & Carlton, P. (2020). Race and work in 2020: How Black professional women break barriers and achieve success. Retrieved from https://www.wharton.upenn.edu/story/race-and-work-in-2020-how-black-professional-women-break-barriers-and-achieve-success/
12. Psychology Today. (2023, June). The invisible biases that influence the science of health. Retrieved from https://www.psychologytoday.com/us/blog/talking-about-health/202306/the-invisible-biases-that-influence-the-science-of-health
13. Linos, E., Mobasseri, S., & Roussille, N. (n.d.). Intersectional peer effects at work: The effect of white coworkers on Black women's careers. Retrieved from https://www.hks.harvard.edu/publications/intersectional-peer-effects-work-effect-white-coworkers-black-womens-careers
14. PubMed. (n.d.). The weathering hypothesis as an explanation for racial disparities in health: A systematic review. Retrieved from https://pubmed.ncbi.nlm.nih.gov/30987864/
15. Harvard T.H. Chan School of Public Health. (n.d.). How discrimination can harm Black women's health. Retrieved from https://www.hsph.harvard.edu/news/hsph-in-the-news/discrimination-black-womens-health/
16. ScienceDaily. (2022, July). Racial discrimination affects brain microstructure, study finds. Retrieved from https://www.sciencedaily.com/releases/2022/07/220720150526.htm
17. CDC. (2022, October 6). Gendered racism among women of color. Retrieved from https://blogs.cdc.gov/healthequity/2022/10/06/gendered-racism-among-women-of-color/
18. Britannica. (n.d.). Wangari Maathai: Biography, nobel peace prize, books, green belt movement, & facts. Retrieved from https://www.britannica.com/biography/Wangari-Maathai
19. This is Africa. (n.d.). Prof Francisca Okeke - Nigerian physicist helping us better understand climate change. Retrieved from https://thisisafrica.me/africans-rising/prof-nneka-okeke-nigerian-physicist-climate-change/
20. Africa Lifestyle. (n.d.). Professor Francisca Nneka Okeke: Pioneering African scientist revolutionizing the world with STEM innovations. Retrieved from https://africallifestyle.com/lifestyle/professor-francisca-nneka-okeke-pioneering-african-scientist-revolutionizing-the-world-with-stem-innovations/
21. UNAIDS. (n.d.). Winnie Byanyima, UNAIDS executive director. Retrieved from https://www.unaids.org/en/aboutunaids/unaidsleadership/bios/winnie-byanyima
22. New Statesman. (2022, January). Winnie Byanyima Q&A: "I'm a nomad in the way I have lived and worked." Retrieved from https://www.newstatesman.com/culture/qa/2022/01/winnie-byanyima-qa-im-a-nomad-in-the-way-i-have-lived-and-worked
23. Poet of Code. (n.d.). Retrieved from https://poetofcode.com/
24. TechCabal. (2023, November 8). This 25-year-old converted her love for cars into a thriving mobility tech startup. Retrieved from https://techcabal.com/2023/11/08/this-25-year-old-converted-her-love-for-cars-into-a-thriving-mobility-tech-startup/
25. AfroTech. (n.d.). Nigerian-American medical student Angela Udongwo is teaching radiologists how to interpret X-rays of patients with Black hairstyles. Retrieved from https://afrotech.com/temple-medical-student-black-hairstyles-research-project

26. Women in Science Africa. (n.d.). Africa has highest proportion of female STEM graduates. Retrieved from https://www.womeninscience.africa/africa-has-highest-proportion-of-female-stem-graduates/#google_vignette
27. Global Energy Alliance for People and Planet. (n.d.). She-STEM potential for a sustainable and inclusive tech-driven economy in Africa. Retrieved from https://energyalliance.org/unlocking-she-stem-potential-for-a-sustainable-and-inclusive-tech-driven-economy-in-africa/
28. Global Voice Group. (n.d.). Women in STEM in Africa: A shift in the social paradigm. Retrieved from https://www.globalvoicegroup.com/news-article/international-day-of-women-and-girls-in-science/
29. UNESCO. (n.d.). Women in science - 2019. Retrieved from https://uis.unesco.org/sites/default/files/documents/fs55-women-in-science-2019-en.pdf
30. Afridigest. (n.d.). 27 women founder-CEOs raised $3 million in Africa. Retrieved from https://afridigest.com/27-women-founder-ceos-raised-3-million-africa/
31. TechSafari. (n.d.). Africa's female founders are crushing it. Retrieved from https://techsafari.beehiiv.com/p/africas-female-founders-crushing
32. ApolloTechnical. (2024). 15 networking statistics everyone should know. Retrieved from https://www.apollotechnical.com/networking-statistics/
33. Harvard Business Review. (2024, March). Research: How women can build high-status networks. Retrieved from https://hbr.org/2024/03/research-how-women-can-build-high-status-networks

2 Centering the History and Progression of the Women's Rights Movement and Its Influence on the Evolving Role of Women in STEM

Pretty Mubaiwa

INTRODUCTION: TELLING A STORY OF ERASURE AND INVISIBILITY: WOMEN IN STEM

In October 2024, the Nobel Assembly at Karolinska Institutet decided to jointly award the 2024 Nobel Prize in Physiology (Medicine) to Victor Ambros and Gary Ruvkun for, "the discovery of microRNA and its role in post transcriptional gene regulation."[1] In the morning when he received the news, the Nobel Prize organization also posted:

> Congratulations to our 2024 laureate Victor Ambros. This morning he celebrated the news of his prize with his colleague and wife Rosalind "Candy" Lee, who was also the first author on the 1993 Cell paper cited by the Nobel Committee.

This post sparked a media frenzy where a number of people questioned why his wife, Rosalind, who co-authored the paper "Cell" did not receive the prize along with her husband and Ruvkun. Rosalind had been a key contributor, starting out as a lab assistant with her husband and co-authoring their first paper "Cell" in 1993. Although it was later clarified that, even though Rosalind was listed as the first author, academic practice dictates that it is the last author who is typically recognized as the leader of the research. This practice, however, should not take away the contribution and needed acknowledgement of Rosalind. To date, a number of researchers who have been active in the field of mRNA are still unaware of Rosalind Lee and her immense contributions.[2]

This incident sparked an even deeper introspective discussion about the history of the Nobel Prize and its erasure of women's contributions. In 1962, Rosalind Franklin's case is another notable instance of the erasure of women and their contribution to

 DOI: 10.1201/9781003540618-2

STEM. In this co-authored and pioneering research by Francis Crick, James Watson and Rosalind Franklin, they published an article about how children inherit their parents' DNA, which was a visionary idea on genetic information that would be used in science and medicine as foundational knowledge. Whilst the researchers worked as a team, Rosalind Franklin provided key discovery and inputs into the research in 1951, when she disproved Watson and Crick's first attempt at creating a DNA model.[3] Although Watson and Crick did acknowledge the contribution of their colleague Rosalind Franklin, her contributions have been largely ignored and/or erased. When the Nobel Prize was awarded, there was no mention or acknowledgement of Rosalind's contributions.

We cannot speak of the erasure of women in STEM globally without telling the story of Marie Curie. In 1902, Marie Curie was nominated and passed over as a co-recipient of the Nobel Prize, where she had been nominated for her innovative and pioneering work in radioactivity along with her co-authors, Pierre Curie and Henri Becquerel.[4] In 1903, the committee renominated Pierre and Becquerel for this similar pioneering work in radioactivity, and this time, no mention was made of Marie Curie. It was clear to the researchers and/or scientists that Marie had discovered radioactivity and single-handedly isolated rhodium as an element.[5] It was Pierre who insisted on the recognition and award of Marie's contribution and efforts. The Nobel Committee responded by awarding Becquerel 70,000 gold francs and having Marie and Pierre share the same amount.[6] This experience, although Marie never fought it, showed how little women were regarded as scientists or worthy contributors to pioneering and innovative knowledge. It is also notable for this discussion that not only is there erasure of women in STEM but also of African scientists in STEM by the Nobel Assembly.

Another interesting story of the erasure of women in STEM is the story of Emmie Warren Roebling, an engineer who contributed immensely to the Brooklyn Bridge's architecture and engineering. At the time of its construction, it was a marvel and an exceptional innovation in architecture. During the construction of the bridge between 1869–1883, her husband, Augustus Roebling, who had taken over the construction from his father, suffered a debilitating illness.[7] This led to Emmie being largely responsible for the construction of the bridge. At that period in history, it was believed that if her name were to be included amongst the engineers of the Brooklyn Bridge, the public would not have had faith in the stability and construction of the bridge. Furthermore, in an article published by the *New York Times*, there was a statement that said, "How the Wife of the Brooklyn Bridge Engineer Has Assisted Her Husband."[8] In some ways, this language overlooked Emmie's pivotal contribution and diminished her role in the construction of the bridge. In this case again, she and her immense contribution to civil engineering were erased. These stories, although not specific to Africa, depict the continued erasure of the contribution of women in the field of science, technology, engineering and mathematics (hereafter, STEM) at some of the highest levels of recognition globally, such as, the Nobel Prize.

Finally, and most importantly related to the subject matter of this book, *African Women in STEM*, it is important to centre African women's history, stories and how they are achieving excellence in a field that was not created for them. According to Brasselman (2003), Western ideologies of patriarchy believed that the female gender does not possess the same aptitude and capability as the male gender, and racial dynamics only added to the bias and thus exclusion of both black men and women.[9] There are a great many untold stories of erasure of African women's contributions to the field of STEM. The story of Fatima al-Fihri is one that elicits great pride for women's contributions to education as a whole and specifically STEM in Africa. She established the University of al-Qarawyyin in 859 AD, and it is the oldest known university on the continent. Established originally as a mosque, the university became the first to award degrees in Islamic studies, mathematics, grammar and medicine.[10] Her contribution to education and the fields of STEM in Africa is notable and should be entrenched in history as a genesis of women in STEM on the continent.

In a paper written by Okune and Mutuku (who are esteemed Techpreneurs in their own right) on becoming an African Techpreneur, they emphasized a critical point that is applicable to African women's experiences in the field of STEM and the erasure and glass ceiling imposed on their critical contributions to the growth of industry in Africa.[11] What is interesting about this form of erasure or non-recognition are the neo-colonial impositions of what is deemed STEM-worthy knowledge, based on imperial and neoliberal constructions of ontology in techpreneurism. Okune and Mutuku note that, "Kenyans figured as Techpreneurs have contested the narrow construction of its parameters (techprenuership), which ironically appear to disproportionately benefit non-Africans working in the Kenyan tech sector."[12] There is an imperial connotation, albeit that is lived by Africans in STEM and particularly women, whereby their contributions are not recognized at the same levels as those of their Western counterparts, and more so, the construction of their contributions is narrowed, and they do not receive the same opportunities as their Western counterparts even in Africa.[13]

These stories above only highlight some mainstream examples of how women in STEM continue to be mistreated, with their contributions systemically and erroneously erased. They continue to face pervasive violence in the form of non-recognition of their contributions, unequal pay and the continued need to break the glass ceiling in the sector. In Africa, particularly, STEM fields have been male-dominated and actively discriminatory towards women through practices such as unequal consideration for female applicants in educational opportunities, discriminatory working conditions and unequal remuneration, to name a few. Women continue to face stereotyping and gender-related biases that impact their educational choices, career journeys and, inherently, their contributions to knowledge, value and impact. Mkhize (2023) surmises this experience very well, stating that, "The colonial ideology designated Africans as subhuman, inferior intellectually, socially, and culturally to the white masculine norm in STEM disciplines."[14] He adds that STEM careers were never created to attract people of colour, let alone women, and this fact contributes to the "inhospitable" nature of the field.[15] This chapter critically explores and examines the historical progression of the global feminist and African women's rights movements and how they have positively continued to influence the participation, inclusion and excellence of African women in the STEM field. The chapter also focuses on pivotal women and human rights frameworks that have been instrumental in raising the flag and helping to create opportunities for entry and equality in all fields of work, but specifically to this book, the field of STEM. Finally, the author provides some recommendations for sustained inclusion and success in empowering women in STEM fields.

UNDERSTANDING THE HISTORICAL PROGRESSION OF THE FEMINIST MOVEMENT

The historical progression of the women's rights movement is a centuries-old one that still bears incremental and positive change for women's lives around the world today. According to Human Rights Watch, "Few movements have made so large an impact in so short a time as the women's human rights movement...[it has] to a large extent also transformed the way human rights issues are understood and investigated."[16] Although over the past few years, one can argue that some of these strides are being eroded through the rise and populism of more right-wing and conservative movements, the women's movement continues to contribute to the inclusion and participation of women in leadership of industries that were deemed "male dominated" such as the STEM fields.

From the days of the first wave of the feminist movement, which culminated in white women in the United States of America obtaining suffrage (the right to vote), although exclusionary, cemented the fight for women to be free to participate as agents, albeit partially, in their lives.[17] The second wave of feminism, beginning in the 1960s, took women's rights to the next level by pushing for the recognition of women's rights, which included the right to work and protections within the workplace to eliminate discrimination and pervasive sexual harassment.[18] This progression led to the establishment of legal reforms to protect women and to provide opportunities to promote equity. Some of the pivotal reforms include the Equal Pay Act of 1963, the Roe v. Wade judgement cementing women's right to have an abortion, and the Civil Rights Act of 1964. This wave also included the beginning of the drafting of the Convention on the Elimination of Discrimination Against Women (CEDAW), which came into force in 1979. Furthermore, as the feminist and women's rights movements have continued to progress, including taking a stand against the violent nature of capitalist institutions and the regulatory frameworks that uphold them against women (in the second and third waves),[19] women in the present day have continued to break barriers to participating in and taking up leadership roles in the STEM field. In the current fourth wave of feminism, the movement continues to hold the mantle, calling for equitable representation in positions of power and decision-making, the right to be free from all forms of violence and discrimination, including the freedom of expression and speaking out against violent socio-normative and cultural practices such as rape culture and sexual harassment.

Again, although significant progress has been made by these waves of feminism, there is much that still remains to ensure gender equality thrives and is cemented as a non-derogable practice to protect women the world over, ensuring they experience and exercise the right to choose their careers, have power over their productive and reproductive capacities, live lives free from all forms of violence and discrimination and are represented and/or have a voice that is heard at all levels of decision-making and power.

BUILDING THE FOUNDATION: THE PROGRESSION OF INTERNATIONAL INSTRUMENTS PROTECTING WOMEN'S RIGHTS

In order to understand the contribution of both the feminist and women's human rights movements to promoting the participation and thriving of women in STEM, it is important to give a short overview of women's rights entrenched in several non-binding and legally binding instruments, both internationally and in the African region. Specifically, the rights to work, equal compensation, education and workplace protection from discrimination and harassment will be highlighted. Human rights law discourse has evolved over time since the adoption of the Universal Declaration of Human Rights (UDHR) by the United Nations (UN) General Assembly in 1948. The Declaration served as the backbone of various other instruments protecting and safeguarding critical and specific human rights, including civil and political, socio-economic, refugee, disability, children, as well as women's rights. In Article 1, the UDHR provisions that all human beings are born free and equal in dignity and rights, thereby providing the fundamental right to equality of all human beings.[20] Specific to women's rights, Articles 16 on marriage, Article 23 on the right to work without discrimination and pay equity/remuneration for equal work, Article 25 on the right to health, which specifically calls out the entitlements of mothers and children to special care and assistance, and Article 26 on the right to education and the right to choose the type of education are all central to cementing the rights of women to participate in any field of work, including those that were traditionally exclusionary to women, such as STEM.[21]

Within the international system of human rights, there are also other instruments specific to women, such as the 1979 Convention on the Elimination of Discrimination against Women (CEDAW), which is known as the International Bill of Rights for Women. The CEDAW is a specific, non-binding instrument that establishes the protection of women from all forms of discrimination based on gender, including in the workplace. CEDAW is the first global treaty that specifies women's rights as human rights.[22] The purpose of CEDAW is to define global standards for defining discrimination against women and encourages states to take steps to eliminate all forms of discrimination against women, as they inhibit the realization of women's core rights.[23] Law-making is a critical step to meeting women's strategic needs because the law provides a means of their realization and an avenue for seeking justice.[24] Through the Optional Protocol to CEDAW, states that have ratified the treaty now have a legal obligation to ensure that the spirit and obligations of the CEDAW are both recognized and given life by giving women, organizations and member states an option to seek and receive compensation for the infringement of their rights as stipulated by the CEDAW.

CEDAW in Article 1 defines discrimination as:

> any distinction, exclusion or restriction made on the basis of sex which has the effect or purpose of impairing or nullifying the recognition, enjoyment or exercise by women, irrespective of their marital status, on a basis of equality of men and women, of human rights and fundamental freedoms in the political, economic, social, cultural, civil or any other field.[25]

This article explicitly calls out the non-derogable nature of discrimination based on sex that impedes women's enjoyment of rights and freedoms afforded to men. States are encouraged to repeal any and all policies, laws and legislation that are discriminatory against women without delay (Article 2), and to adopt gender empowering and equity-appropriate legislation and policies. Article 2 also outlines positive and negative obligations on the state to ensure that member states take steps to ensure the equality of men and women, which can also be interpreted to bear positive value for women in education, the workplace and any other field where they act as agents. Articles 3 and 4 can also be related to the advancement of women in STEM as the state is encouraged to ensure the development and advancement of women in all fields, as well as the adoption of special measures to accelerate equality between women and men. By imposing this obligation on states, this has created and affirmed the responsibility of states to ensure that there is the creation of opportunities, including undertaking special measures such as affirmative action to encourage the participation of women in STEM.

There are also pivotal human rights instruments which have played a role in ensuring the inclusion of women in the workplace and vitalizing their right to career choice and empowerment in the field of STEM. These include the 1981 International Labour Organisation (hereafter ILO) Workers with Family Responsibilities Convention, which aims to promote equal treatment for men and women workers.[26] The Maternity Protection Convention, 2000, regarded as the most forward-thinking framework for women in the workplace, established women's maternity protections such as the right to take leave from work for maternity purposes, receive maternity benefits, employment protection and breastfeeding.[27] There are several other instruments protecting women's participation in education and the workforce that have supported women in all fields of education and work, including STEM, without which discriminatory practices, exclusion and erasure would continue to this day.

Despite these developments in human rights law, none of the instruments stated above are legally binding on states. These treaties and frameworks need to be adopted by states in their municipal law for women, girls and other disadvantaged groups to access justice using their domestic judicial systems. Moreover, there are requirements for the exhaustion of domestic remedies before accessing regional or international justice systems, which further impacts the ability of women to claim their rights in this space and others. The lack of legally binding instruments thus creates normative gaps and legal vacuums that negatively influence the positive empowerment, equity, recognition and participation of women because there is no "reasonably" accessible recourse for women who have faced discrimination regarding their choice to get a STEM education, be employed equitably in the field, as well as to receive equal compensation as their male counterparts in the same field. Additionally, due to the rise and populism of right-wing and conservative politics and movements in Western politics, the progress that has been made is being eroded. One of the seminal points is the overturning of *Roe v. Wade* in 2022 by the ruling in *Dobbs v. Jackson Women's Health Organisation*. This judgement clawed back the rights of women to choose to abort a pregnancy. As a norm creator and leader, having women's rights being eroded in the US poses a threat to the rights of women worldwide.

BRINGING IT HOME: THE PROGRESSION OF WOMEN'S RIGHTS MOVEMENT THAT ENABLE THE INCLUSION OF WOMEN IN STEM IN AFRICA

Women in Africa suffer from cultural, political and socio-economic barriers to the realization of their rights. In comparison to other parts of the world, African women's conditions of life continue to deteriorate despite the progression of different human rights systems both regionally and internationally.[28] Women in Africa face a daily struggle for survival that limits their experience and appreciation of human rights due to the systemic and structural environments in which they live. Gender equality and women's empowerment in this respect remain a critical and fundamental area of concern for African women in general; more so, for the marginalized, minority and rural women. It is important to ensure that any contemporary discussion about Africa includes an intersectional analysis that takes into consideration the continent's history of oppression under colonialism and how this history continues to have egregious and lasting effects on equality and the progression of women. Africa's experience of colonialism reflected a special type of oppression and dominance,[29] to speak of any development in African history without making reference to this important period of history would be an injustice to any subject matter, including the progression and contribution of women in STEM. This is because colonialism created a transmutation of social organization, participations and expression that still affects contemporary sociopolitical and economic context.[30]

Human rights can be traced back to African communities' pre-colonial period through the different collective structures which were established to secure and assure the happiness of all members of the society. Colonialism transmuted human rights as they took a new form when the colonial state enforced a new system of rights and justice which was exclusionary as it created two rights groups of citizens and subjects.[31] The brutal denial of rights contributed to the stifling of democracy in African communities.[32] The setup of the colonial administration denied many people their dignity, and this colonial administration carried out ruthless enforcement which indicated a disregard for the rights and freedoms of people, including women and their right to education and work.[33] The colonial authorities created a bifurcated state that established categories of citizenship, which determined the rights awarded to each group of people. In a nutshell, the colonial period imposed a regression on the development of human rights on the continent.[34] Banda argued that women's rights were even more affected during colonialism because, "there was a selective understanding of legal norms by male colonial officers who accorded their women fewer rights than those enjoyed by men" and thus women were largely excluded from norm making.[35] This discrimination contributed to the inequality of access to education, with women not being recognized as equal contributors who are able to make positive contributions, including in the STEM field.

Africa underwent a period of decolonization from the late 1950s to late 1980s, and the pivotal role of the newly independent states became to re-establish citizenship which included re-establishing human rights for everyone.[36] The Organization of African Unity (OAU) was established at the regional level and was the first regional cooperation mechanism on the continent. However, the Charter to the OAU did not make any commitment to protect women's rights and even more so achieving gender equality and empowerment.[37] The OAU's main concern was the decolonization of the continent whilst maintaining sovereignty and integrity of the newly formed states.[38] Therefore, during this period, human rights were not the primary concern. In 1981, external pressures mounted due to increasing gross human rights violations on the continent through civil wars and the rise of oppressive/dictatorial regimes. The OAU adopted the African Charter, and this is how the African human rights system was established. The African Charter on Human and People's Rights was also meant to address some criticism about the universality and relativity of the international human rights instruments to the African context. The African Charter was established as a contextually relevant human rights framework that catered to Africa's unique histories and contexts. According to Chirwa, the African Charter therefore is distinctive in that it includes an "African fingerprint" to human rights.[39] The Charter also distinctively included provisions for women's rights in Articles 2 (non-discrimination on the basis of sex) and 18 (3) (state obligation to ensure the elimination of every discrimination against women).

The African Charter was criticized for focusing more on community and group rights and largely ignoring the rights of women and children.[40] The African Charter has a communitarian focus that completely ignored individual rights, ergo, women's individual rights, and this placed women's rights in a legal comma.[41] "The African Charter also reinforces gender stereotypes that 'tolerate, confirm and support repressive structures' of traditional African societies to the detriment of women's rights."[42] On a positive note, specifically related to the right to education, work, equality and empowerment, the African Charter in Articles 2 and 19 reinforces the rights to non-discrimination on various grounds, including sex and equality, respectively. However, on the other hand, the Charter gives with one hand and takes with the other, because throughout the framework, there is a continuous reiteration of the respect for cultural practices, many of which do not view women as equal contributors to economic, political and family life.[43] This contradiction is detrimental to the realization of gender equality and women's empowerment in Africa. The dualism of the Charter presents both a normative and interpretive gap, as culture reinforces elements of discrimination between sexes including differing gender roles and responsibilities, which impacts the positive inclusion and recognition of women in the fields of STEM. On the other hand, principles of non-discrimination and equality mean that under no circumstances can any discrimination be made on the basis of gender or sex, which continues to be an oxymoron in the fight for equality and empowerment for women in the field of STEM in Africa.

The adoption of the Protocol to the African Charter on Human and Peoples' Rights on the Rights of Women in Africa (also known as the Maputo Protocol) in 2003 and its entry into force in 2005 came as a welcome addition to the promotion and protection of women's rights on the continent. In its Preamble, member states recognized the crucial role of women in the preservation of African values and that despite member states' ratification of the African Charter, women still continued to be victims of discrimination and harmful practices.[44] The Women's Protocol imposes obligations on the state to respect, protect, promote and fulfil different women's human rights in their jurisdictions. The Maputo Protocol provides an avenue for women to seek justice if they face any form of discrimination, whether in public or private life, including in their education or their workplace. This is an important instrument that protects women's rights to education, work, contribute to knowledge and participate in the field of STEM. Without this fundamental basis, women would continue to face oppression, and exclusion in the fields of STEM, whilst at the same time not having a means for recourse for injustice.

Outside of the normative frameworks, the African Union's Agenda 2063, under Aspiration 1 for a prosperous Africa based on inclusive growth and sustainable development, para 10 specifically calls out the mission, "*to create a prosperous continent that...has well educated citizens, underpinned by science, technology and innovation*"[45] Additionally, under Aspiration 6 focused on women, youth and children, para 58 specifically calls out young African men and women being supported as path breakers of African knowledge and society, contributing to innovation and entrepreneurship.[46] This can be interpreted and used to advocate for building and investing in STEM fields. In 2015, the AU proclaimed 2015 the Year of Women's Empowerment and Development, which shows both a recognition of and commitment to addressing and closing the gaps in women's empowerment generally. Additional frameworks that specifically aim to empower women in STEM include the AU's Science, Technology, and Innovation Strategy for Africa (STISA-2024), which aims to boost the participation of women in science, technology and innovation enabled socio-economic development. Other bodies such as the African Union High-Level Panel on Emerging Technologies (APET) have been instrumental in advising member states to create enabling environments for women in STEM, establish STEM empowerment-related policies including labour policies that promote work-life balance for STEM personnel in general, incorporate gender equity into STEM education by empowering STEM female teachers, encourage enrolment of young women and girls into STEM fields and to provide the means for realization.[47] The initiatives and frameworks highlight a few of the activities taking place on the continent from a policy perspective. It is also important to draw the nexus between the continued development of these policies, initiatives and programmes with the foundations of women's rights that are rooted in the African human rights system, which shapes and encourages member states and stakeholders to affirm and empower the position of women generally. Without these fundamental foundational frameworks continuing to evolve and push for change, related to ensuing global and regional feminist waves which seek to create an equitable and empowered future for all women.

SUSTAINING THE MOVEMENT FOR FUTURE GENERATIONS: HOW DO WE MAINTAIN THE GAINS?

In 2020, we witnessed the global protest called #ShutDownStem. This was a protest by black scientists against racism and exclusion in science and academia. This protest was not based on gender discrimination; it was centred around demanding transformation for black people in STEM. This protest in 2020 showed the continued criticality of transformation in the field. It stands to reason that if the representation of an entire racial group is dismal, then the representation of black women is even worse, and the same can be assumed for African women. Scholars and practitioners continue to opine that there is a chronic underrepresentation of women, which is exacerbated by the STEM fields remaining largely untransformed.[48] According to the US Bureau of Labor Statistics, Women's Bureau, from 1970 to 2020, women occupying STEM positions in the United States had only grown from 7 to 26 percent.[49] The Pew Research in 2021 also released some glaring statistics which also confirmed the underrepresentation of women in STEM fields generally. Their research established, for example, that women make up a majority of workers in the healthcare industry but remain underrepresented in other job clusters such as the physical sciences, computing and engineering.[50] This research also noted that, whilst women were enrolling and graduating with a majority of undergraduate and graduate degrees, they remained a small share in earning STEM related degrees.[51] Although there has been an increase in enrolment, there are still structural and systemic changes that need to be made to holistically empower women in education and the workplace in the STEM field. Research has also shown that African universities do not have gender equality and discrimination policies,[52] this further impacts the throughput of graduates into the market because systemic biases and discrimination affect how women are able to study and graduate if the system does not have enablers for them to be successful.

Across the African continent, governments, investors, universities and other stakeholders have continued to take positive steps to attract women to the fields of STEM, including applying quotas for enrolment, offering scholarships and other equity-enabling practices. A report compiled by Times Higher Education (THE) and the United Nations Scientific Cultural and Educational Organisation established that 47 percent of graduates in STEM in Africa, at both undergraduate and postgraduate levels, are women.[53] The World Bank statistics have also confirmed that African universities have the highest percentages of women graduating in sciences.[54] Despite enrolment and graduate numbers for African women increasing, a deeper introspection of the numbers shows several gaps in establishing equity and empowerment for women in STEM. These gaps include limited gender equity and empowerment policies in universities, insufficient mentoring practices or schemes for women and the provision of scholarships.[55] Specific to Sub-Saharan Africa, the number of women graduating from engineering-related degrees falls below 30 percent.[56] Again, this shows the existence of gaps at the level of implementation, which should be addressed in order to maintain and increase the numbers across different STEM fields over time. Some scholars such as Borum and Walker (2012) argued that even though the number of women enroling in the STEM fields is increasing, there is a wider gap for sustained enrolment in mathematics.[57] However, women still remained underrepresented at staff and academic levels.[58]

Innes (2022) states that there is a leaky pipeline that starts even before enrolling at the tertiary level. Young women and girls in Namibia, and South Africa, for example, have the highest secondary enrolments; however, many of them drop out and/or do not complete secondary education, thereby diminishing their chances for enroling and earning an undergraduate or graduate degree in STEM.[59] Women's lives exist on a spectrum according to feminist social theory. Therefore, their progression or lack thereof in secondary school, and even at undergraduate and postgraduate levels, is impacted by several other factors that affect their daily lives, such as cultural biases, multiple forms of violence and discrimination (also related to teenage pregnancy and mortality rates), economic barriers and environmental barriers such as lack of political will to improve the position of young women and girls throughout high school and undergraduate levels. According to the Equality Equation Report, gender biases, stereotypes, gender norms and sexual harassment are key drivers for the continued underrepresentation of women in STEM.[60]

CONCLUSION

The drivers and effects of the underrepresentation of women in STEM are known. This underrepresentation is not a unilinear phenomenon; it starts in the very context of the family where there is a low value placed on girl children, influenced by age-old cultural normative standards and belief systems. This cycle of invisibility, erasure and a lack of value continues with the choices parents make on whom to send to school. Girl children bear the brunt of the economic difficulties experienced by their parents and drop out of school because their parents or guardians believe that sending the boy child, when there are limited resources, will benefit the family in the longer term more than sending the girl child. Furthermore, women and girls experience multiple and intersecting forms of violence in their familial and community spheres that hinder them from pursuing a successful education and career in STEM, as argued above. Even when they manage to break those barriers in education and earn a STEM qualification in the workspace, women continue to experience discrimination and violence that inhibit their growth, learning, progression into leadership and equal remuneration for equal work.

Therefore, the author recommends adopting an intersectional lens to ensure the holistic empowerment of women in STEM. It is not enough to start trying to encourage young women and girls in secondary and tertiary education to enrol for STEM qualifications. Whilst these practices are pivotal to progress, it is also important to push for continued laws and policies that promote the advancement of women and girls in education as a whole. For the STEM specific field, these laws and policies create the backbone for social and behavioural change, as well as state obligations that will encourage change at multiple levels where women's lives exist and they act as agents. Secondly, it is recommended that continued advocacy for women in leadership and decision-making is critical to ensure sustainable change. Research has shown that the inclusion of women in leadership (in any field) improves the success, equity and delivery of impact in different fields.

NOTES

1. Press release, The Nobel Prize in Physiology or Medicine 2024 available at https://www.nobelprize.org/prizes/medicine/2024/press-release/ accessed on 10 October 2024.
2. Press release, The Nobel Prize in Physiology or Medicine 2024 available at https://www.nobelprize.org/prizes/medicine/2024/press-release/ accessed on 10 October 2024.
3. El Pais, How the secret of life was discovered 70 years ago, available at https://english.elpais.com/science-tech/2023-04-25/dna-how-the-secret-of-life-was-discovered-70-years-ago.html accessed on 10 October 2024.
4. https://lithub.com/when-marie-curie-was-almost-excluded-from-winning-the-nobel-prize/#:~:text=Following%20Nobel's%20death%2C%20the%20first,because%20she%20was%20a%20woman.
5. https://lithub.com/when-marie-curie-was-almost-excluded-from-winning-the-nobel-prize/#:~:text=Following%20Nobel's%20death%2C%20the%20first,because%20she%20was%20a%20woman.
6. https://lithub.com/when-marie-curie-was-almost-excluded-from-winning-the-nobel-prize/#:~:text=Following%20Nobel's%20death%2C%20the%20first,because%20she%20was%20a%20woman.
7. Brittanica 'Emmie Warren Roebling: American Socialite, builder and business-woman article' available at https://www.britannica.com/biography/Emily-Warren-Roebling.
8. https://timesmachine.nytimes.com/timesmachine/1883/05/23/102822831.html?pageNumber=1.
9. Braselmann (2003)
10. https://theglobalacademy.ac/the-worlds-first-university-was-founded-by-a-woman/.
11. Okune and Mutuku (2023).
12. Braselmann (2003).
13. Press release, The Nobel Prize in Physiology or Medicine 2024 available at https://www.nobelprize.org/prizes/medicine/2024/press-release/ accessed on 10 October 2024.
14. Mkhize (2023)
15. Mkhize (2023).
16. Human Rights Watch (1995).
17. Tanja (2024).
18. https://timesmachine.nytimes.com/timesmachine/1883/05/23/102822831.html?pageNumber=1
19. https://timesmachine.nytimes.com/timesmachine/1883/05/23/102822831.html?pageNumber=1
20. Article 1 of the Universal Declaration of Human Rights (UDHR), 1948.
21. Articles 16,23, 25 and 26 of the UDHR.
22. Ulrich (2000).
23. Article 1 of the Convention on the Elimination of All Forms of Discrimination Against Women.
24. Cook (1995).
25. Article 1 of the Convention on the Elimination of Discrimination Against Women, 1979.

26. 1981 International Labour Organisation Workers with Family Responsibilities Convention No. (156).
27. 2000 Maternity Protection Convention No. (183).
28. Welch (1993).
29. Fanon (1963).
30. Oloka-Onyango (2000).
31. Mamdani (1996).
32. Kunig et al. (1985).
33. https://lithub.com/when-marie-curie-was-almost-excluded-from-winning-the-nobel-prize/#:~:text=Following%20Nobel's%20death%2C%20the%20first,because%20she%20was%20a%20woman.
34. Mubaiwa (2021).
35. Banda (2006).
36. Mubaiwa (2021).
37. Chirwa (2006).
38. Chirwa (2006).
39. Chirwa (2006).
40. Human Rights Council Report of the Special Rapporteur on violence against women, its causes and consequences, Rashida Manjoo (2015) at 5.
41. Elmadmad and Alloo (1992).
42. M wa Mutua (2021).
43. See Articles 17 (2) and 29 (7) of the African Charter
44. Para 8 Preamble of Women's Protocol states that 'Reaffirming that the principle of promoting gender equality as enshrined in the constitutive Act of the African Union as well as the New Partnership for Africa's Development, relevant Declarations, Resolutions and Decisions which underline the commitment of the African States to ensure the full participation of African women as equal partners in Africa's development...'
45. Africa Union 'Agenda 2063: The Africa we want' available at https://au.int/en/agenda2063/overview.
46. Africa Union 'Agenda 2063: The Africa we want' available at https://au.int/en/agenda2063/overview.
47. Heightening the participation of African Women in Science, Technology, Engineering and Mathematics Career Paths, article available at https://www.nepad.org/blog/heightening-participation-of-african-women-science-technology-engineering-and-mathematics
48. Mkhize and Idahosa (2021).
49. United States Department of Labour available at https://www.dol.gov/agencies/wb/data/occupations-stem accessed on 18 October 2024.
50. Fry et al. (2021).
51. Fry et al. (2021).
52. World Bank Group (2020).
53. Times Higher Education (2022).
54. Marie Francoise Marie-Nelly blog titled Tackling bias on this International Women's Day (2022) available at https://blogs.worldbank.org/en/nasikiliza/tackling-bias-international-womens-day.
55. Alex Rose Innes 'Africa has highest proportion of female STEM graduates' (2022) available at https://www.womeninscience.africa/africa-has-highest-proportion-of-female-stem-graduates/.
56. Alex Rose Innes 'Africa has highest proportion of female STEM graduates' (2022) available at https://www.womeninscience.africa/africa-has-highest-proportion-of-female-stem-graduates/.

57. Borum and Walker (2012).
58. Borum and Walker (2012).
59. Inness Fry et al. (2021).
60. World Bank Group (2020). United States Department of Labour available at https://www.dol.gov/agencies/wb/data/occupations-stem accessed on 18 October 2024.

REFERENCES

2000 Maternity Protection Convention No. (183).

Universal Declaration of Human Rights (UDHR), 1948.

1981 International Labour Organisation Workers with Family Responsibilities Convention No. (156).

1979 Convention on the Elimination of Discrimination Against Women.

Africa Union 'Agenda 2063: The Africa we want' available at https://au.int/en/agenda2063/overview.

Banda, F. (2006). Women, law and human rights in Southern Africa. *Journal of Southern African Studies*, *32*(1), 13–27.

Borum, V., & Walker, E. (2012). What makes the difference? Black women's undergraduate and graduate experience in mathematics. *Journal of Negro Education*, *81*(4), 366–378.

Braselmann, S. (2003). Reluctant rebels: Women scientists organizing. *Bulletin of Science, Technology & Society*, *23*(1), 6–9.

Brittanica 'Emmie Warren Roebling: American Socialite, builder and business-woman article' available at https://www.britannica.com/biography/Emily-Warren-Roebling.

Chirwa, D. M. (2006). Reclaiming (Wo)manity: The merits and demerits of the African protocol on women's rights. *Netherlands International Law Review*, LIII, 63–96.

Cook, R. J. (1995). Enforcing women's rights through law. Gender & *Development*, *3*(2), 8–15.

Elmadmad, K., and Alloo, F. (1992). The human rights of women in the African Charter. In Wolfgang, B., & Wolfgang, H. (Eds.), *Regional systems of Human Rights Protection in Africa, America and Europe*. Friederich Naumann Foundation, Brussels.

El Pais, How the secret of life was discovered 70 years ago, available at https://english.elpais.com/science-tech/2023-04-25/dna-how-the-secret-of-life-was-discovered-70-years-ago.html

Fanon, F. (1963). *The Wretched of the Earth*. Grove Press, New York.

Fry, R., Kennedy, B., and Funk, C. (2021). STEM Jobs See Uneven Progress in Increasing Gender, Racial and Ethnic Diversity: Higher education pipeline suggests long path ahead for increasing diversity, especially in fields like computing and engineering. available at https://www.pewresearch.org/social-trends/2021/04/01/stem-jobs-see-uneven-progress-in-increasing-gender-racial-and-ethnic-diversity/

Human Rights Watch. (1995). The Human Rights Watch Global Report On Women's Human Rights.

Kunig, P., Wolfgang, B., and Mahalu, C. R. (1985). Regional protection of human rights by international law: The emerging African system *Baden-Baden*, Verlagsegesellschaft.

Oloka-Onyango, J. (2000). Human rights and sustainable development in contemporary Africa: A new dawn or retreating Horizons? *Human Rights Law Review*, *39*, 43–51.

Mamdani, M. (1996). *Citizen and Subject: Contemporary Africa and the Legacy of Late Colonialism*. Princeton University Press, Princeton.

Mkhize, Z., (2023). Is it transformation or reform? The lived experiences of African women doctoral students in STEM disciplines in South African universities. *High Education*, *86*, 637–659.

Mkhize, Z., and Idahosa, G. E. (2021). Intersectional experiences of Black South African female doctoral students in STEM: Participation, success and retention. *Agenda Empowering Women for Equity Journal*, 1–13.

Mubaiwa, P., (2021). *Human Rights, Modernity and Culture: Do the current international and regional human rights frameworks and discourse include Lobola and a form of violence against women?* [Doctoral dissertation, University of Cape Town].

Okune, A., and Mutuku, L. (2023). "Becoming an African techpreneur: Geopolitics of investments in "local" kenyan entrepreneurship." *Engaging Science, Technology, and Society*, *9*(1), 81–103.

Tanja, S. (2024). The evolution of feminism: Tracing progress and challenges. *Journal of International Social Research* 17, 108.

Ulrich, J. (2000). Confronting gender-based violence with international instruments: Is a solution to the pandemic within reach?' *Indiana Journal of Global Studies*, *7*(2), Article 11.

wa Mutua, M. (2021). The African human rights system in a comparative perspective: The need for urgent reformulation. *The Nairobi Law Monthly*, 27.

Welch, E. C. (1993). Human rights and African women: A comparison of protection under two major treaties. *Human Rights Quarterly*, *15*, 550.

World Bank Group. (2020). The equality equation: Advancing the participation of women and girls in STEM. The International Bank for Reconstruction and Development/World Bank. available at https://documents1.worldbank.org/curated/en/789951595308672516/pdf/Main-Report.pdf

3 From Classroom to Boardroom
Success Stories of African Women in STEM

Bimpé Olufemi

INTRODUCTION

According to the United Nations Educational, Scientific and Cultural Organization (UNESCO), only 35% of graduates in the STEM fields are women. This statistic has stayed the same for the past decade and is influenced by biases, societal norms, and expectations that affect the quality of education women receive and the field of study they opt for. Women's exclusion from STEM education is harmful not only to their future careers but also to the overall well-being of society, technological progress, economic development, and the progress towards sustainable development goals.

According to recent data from Times Higher Education and UNESCO, 47% of university STEM graduates in Africa are women, the highest percentage worldwide. However, it is painful to note that less than 22% of these gifted women hold leadership positions at the board level.

Dr Francisca Nneka Okeke, a renowned physicist and the first African woman to win the L'Oréal-UNESCO For Women in Science Award, comments: "The high percentage of women STEM graduates in Africa is encouraging, but we must ensure these women have equal opportunities to rise to leadership positions. Only then can we say we've truly succeeded" (Okeke, personal communication, 2023).

Stefania Giannini, UNESCO Assistant Director-General for Education, said, "We need to unlock the potential of STEM education across Africa. African girls represent the greatest untapped population for becoming the next generation of innovators."

UNESCO is confident that by implementing the right pedagogical methodologies, we can increase the number of high school girls passionate about STEM. This will include tailoring the learning styles for girls and making focused lesson plans for them. Additionally, we must address cultural biases, provide financial support, and offer strong role models to inspire and empower these young girls.

Why is it important to groom girls from an early age? Our survey demonstrates that 80% of our women leaders were introduced to STEM in High School (Figure 3.1).

DOI: 10.1201/9781003540618-3

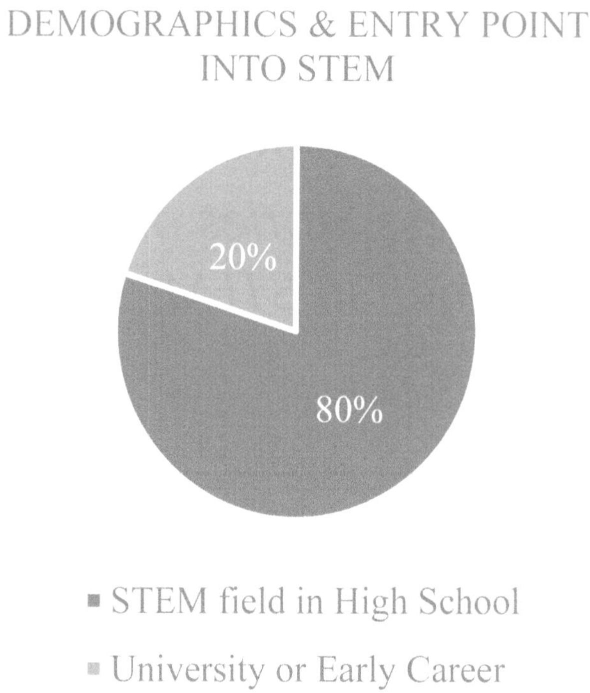

FIGURE 3.1 Introduction to STEM

This chapter showcases the success stories of accomplished female STEM leaders and presents the survey results. The insights from the survey highlight career journeys, factors contributing to success, and recommendations for aspiring STEM leaders.

THE CURRENT LANDSCAPE

Over the last few decades, there has been a notable increase in African women enrolling in STEM courses across Africa. Grassroots movements, government policies, non-governmental organisations, and international support have also influenced the rise. Nevertheless, the achievements made during this period continue to be hampered by several factors that inhibit African women from fully engaging in and benefiting from STEM fields. These challenges include limited access to quality education, cultural and societal barriers, and a lack of representation in leadership roles within STEM fields. There are, however, ongoing efforts and initiatives to tackle these issues and improve the situation for African women in STEM.

Increased Representation

The number of women enrolling in degrees and professional engineering, computer science, and medicine courses has increased, particularly in countries like Nigeria,

South Africa, and Ghana. Also, information available highlights substantial growth in the number of women engineers in Ethiopia. This trend is perceived as a welcoming undertaking for increased gender diversity within STEM fields across Africa.

Moreover, there is growing visibility of women in tech entrepreneurship, highlighting their strength as entrepreneurs and their ability to thrive in gender-inclusive roles. Historically, women on the African continent have always been very enterprising, and the tech field now provides them with the flexibility and autonomy to flourish. Nonetheless, the funding gap places them in a less advantageous position than their peers in high-income countries, hindering their full participation and success in STEM and entrepreneurship.

Policy Support

Over the last couple of years, most African governments have undertaken plans to support women's empowerment and change the status quo of women in society. Countries such as Rwanda and Uganda have been at the forefront of ensuring that differences in boys' and girls' enrolment in schools in science subjects are addressed. These policies include, but are not limited to, the provision of affirmative action scholarships, women in STEM mentorship programmes, and other additional initiatives that aid women in STEM career fields. They have also reinforced such actions in policies such as the African Union Agenda 2063, which has helped many women pursue STEM careers in the various countries on the African continent.

Mentorship and Role Models

Some remarkable African women who are successful STEM professionals include Dr Ngozi Okonjo-Iweala, the DG of the World Trade Organisation, Funke Opeke, CEO and founder of MainOne Cable, Research Chair Dr Tebello Nyokong from Rhodes University, and Farida Bedwei of Logiciel. These women have emerged as exemplary role models, inspiring the younger generation and proving that success in STEM fields is within reach for African women. In addition, they have made significant achievements in their careers and have played an important role in promoting STEM in Africa(Figure 3.2).

Over the past decade, many notable mentorship programmes have emerged to encourage young girls to take up and become adept at STEM subjects. This is a welcome development and aligns with our survey results, which indicated a need for suitable mentors and mentorship programmes. For instance, She Codes Africa (SCA) teaches and attends to young girls interested in technology, while the Women's Technology Empowerment Centre (WTEC) focuses on helping girls and women in Nigeria through mentorship and the provision of networks. Similarly, STEMi Makers of Africa focuses on making it easy for African women in STEM to develop their careers by providing them with mentorship and skill acquisition programmes. Such initiatives offer mentorship and networks, encouraging young African women to pursue STEM careers and access invaluable support to succeed in such fields.

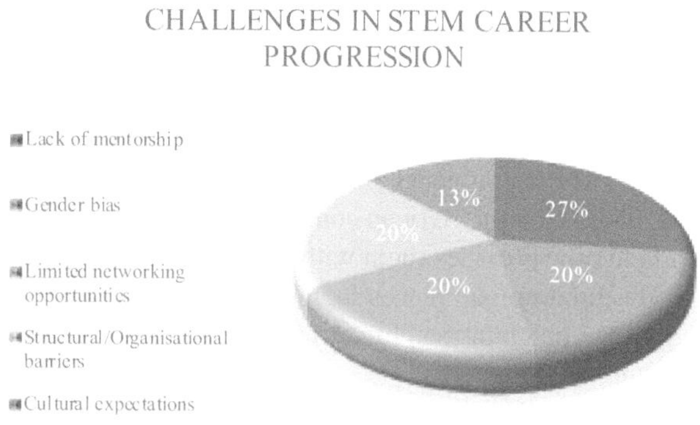

FIGURE 3.2 Challenges in STEM progression

STEM Advocacy and Grassroots Movements

Non-governmental organisations (NGOs) have actively promoted women in tech in several African nations. For instance, AkiraChix, located in Kenya, helps women in STEM by providing training programmes and a supportive environment. Similarly, the WAAW Foundation in Nigeria also focuses on women in tech to provide advocacy and support. Such organisations inspire women to reach out and share their abilities.

Festivals and competitions are great ways to encourage young girls to participate in STEM (especially the gamified ones). For example, Africa Code Week is held annually to teach youths how to code. This initiative promotes digital skills and adds more young women to the STEM pool. The Technovation Challenge provides a platform for girls worldwide, including Africa, to develop mobile applications and start their businesses. The festivals and competitions are instrumental in raising awareness about the importance of women in STEM and providing opportunities for them to demonstrate their technological and innovation skills.

Technological Advancements

The rise of digital platforms like Coursera, LinkedIn Learning, and Funzi in Kenya has provided access to online courses aimed at upgrading the knowledge and skills of African women irrespective of geographical limitations, thus establishing the concept of learning without borders.

Increased Visibility

African women in STEM fields are gaining increasing recognition for their remarkable contributions to science, technology, engineering, and mathematics. Their achievements are celebrated through prestigious awards and extensive media coverage from reputable media channels such as Forbes, CNN, BBC, and TechCabal. One such personality, Dr Ameenah Gurib-Fakim, is a renowned biologist who has also served as the president of Mauritius. Her fame on the international stage shows appreciation for her work, helps combat and reshape gender norms and stereotypes, and inspires African women in STEM.

THE CHALLENGES

Gender Stereotypes and Cultural Biases

Gender stereotypes and cultural biases often create significant barriers for women who are interested in pursuing STEM careers. These biases can manifest in various ways, such as societal expectations that steer young girls away from science and maths or the lack of female role models in STEM professions. Additionally, in rural areas, where over 50% of the African population resides, there are unique challenges like limited access to educational resources, early marriages due to cultural norms, and family responsibilities that hinder young girls from pursuing higher education in STEM.

Limited Access to Quality Education

A shortage of funding for educational resources, inadequate infrastructure such as outdated facilities, insufficient electricity and ventilation, lack of access to technology, and a lack of adequately trained educators with qualifications in STEM contribute significantly to providing poor-quality education in these areas and limit the students' exposure to modern educational tools and resources. These challenges are more common among rural and underserved African school populations where young girls, in general, face significant barriers to accessing quality education. For example, in some circles, it is perceived as more prudent to educate boys than girls due to the future return on investment for the families. However, this thinking is changing due to widespread efforts to create awareness. These obstacles result in a higher number of female student dropouts, perpetuating the cycle of limited educational opportunities.

Funding and Economic Barriers

Financial constraints remain a significant barrier for African women pursuing STEM education. Tuition, textbooks, and other educational fees are usually too expensive, especially for people from underprivileged backgrounds. While some opportunities exist for grants and scholarships, these are generally highly competitive, leaving many potential students out in the cold.

In addition, many STEM initiatives targeting women in Africa are underfunded, limiting their ability to reach and support a large number of women interested in STEM fields. This trend is a significant drawback to the sustainability of the programmes. It continues to undermine their long-term impact, making it challenging for them to provide ongoing support, resources, and mentorship to women pursuing STEM careers. As a result, many promising initiatives struggle to achieve their full potential in addressing the persistent gender gap in STEM fields.

Workplace Discrimination and Gender Bias

Women face gender pay gaps compared to their male counterparts, which become increasingly visible as they climb the career ladder, resulting in the "glass ceiling" effect. While not all women acknowledge the existence of pay gaps, more women recognise this issue than deny it. In addition, due to limited role models at the helm of leadership to push their cause, women in STEM face hiring and promotion biases, further limiting their career advancement. In my leadership capacity, I was instrumental in hiring, promoting, and moving many young women who would otherwise have been overlooked into STEM fields. The right corporate policies, enforced, could make a tremendous positive impact, especially in diversity, equity, and inclusion, not just on a company-wide scale but also at divisional and departmental levels(Figure 3.3).

People hire people like themselves

The STEM workforce in many African countries remains male-dominated, which can be daunting for women with aspirations in this space and can lead to issues like isolation, harassment, and a lack of support networks. It takes an intuitive leader to understand the skills and needs of their workforce to direct their careers positively. As a leader of people, I have supported numerous young women and men in moving in the direction of their passion and navigated openings for these moves to happen. Usually, the most significant challenge for women is the ability to balance work and

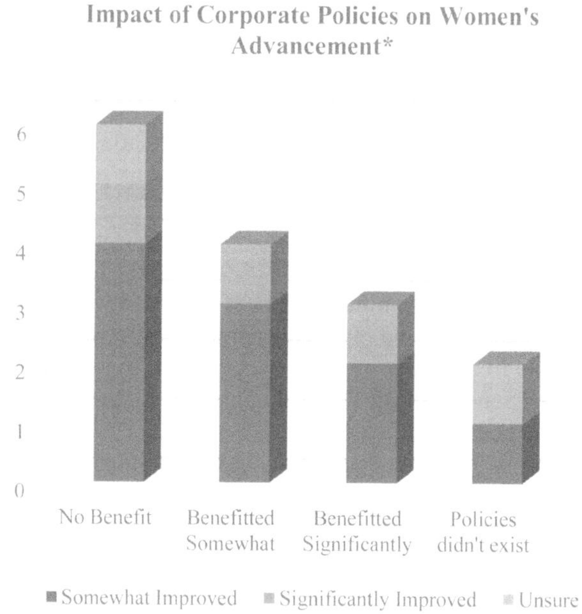

FIGURE 3.3 Stacked bar chart showing perceived landscape changes

the demands of a young family, which can take its toll with a negative impact on their mental health and career progression.

Lack of Mentorship and Networking Opportunities

In rural and deprived communities, women face significant barriers to accessing mentors and mentorship programmes. They also lack skills and access to networking opportunities which can advance their STEM careers. Factors such as limited resources, lack of infrastructure, and geographical barriers contribute to the remoteness of these critical support systems. As a result, extra help is required to enable them to build valuable connections, gain knowledge, and receive guidance from experienced professionals in their field.

In my experience, though some companies attempt to run structured mentorship programmes for women, they fall short because the effort required to sustain them is greater than the available resources to run and commit to them. Men often network through social events at bars, clubs, etc., where mentorship relationships and deals are forged. However, women's ability to participate in such activities at that time and place is limited due to family commitments. This leaves them with business hours to forge networks and deliver stellar work. They are working doubly hard, but this is visible with a negative impact mainly in the family home, not the desired positive impact in the corporate home. Some survey respondents acknowledge that having proper support, whether family or paid help (e.g., housekeeper, nanny, chauffeur, etc.), has helped provide the needed time for their professional activities. Still, paid help is not usually accessible in the early years of one's career while climbing the ladder(Figure 3.4).

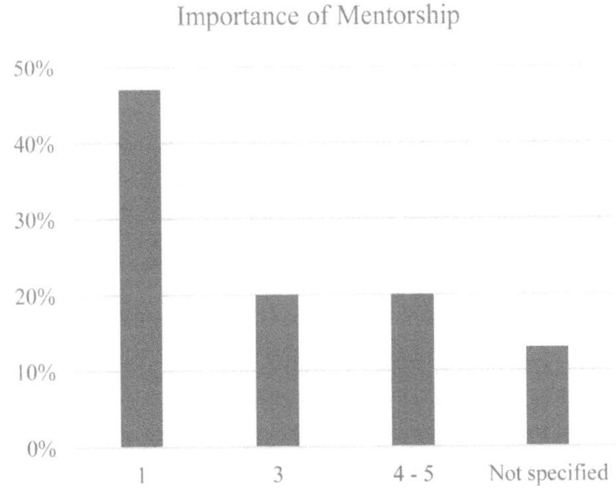

FIGURE 3.4 Distribution of mentorship importance ratings

Technological Divide

Despite the increasing technological advancements, the digital divide continues to pose a significant challenge, especially in rural areas. Limited access to computers, high-speed internet, and STEM-related resources hampers the ability of young women from rural communities to engage in STEM learning and pursue careers in fields such as computer science, engineering, and information technology. This lack of access affects educational opportunities and limits their ability to participate in the digital economy and access online resources for personal and professional development.

Even with the proliferation of digital platforms, there are gaps in technical skills training, particularly in high-tech fields such as artificial intelligence, machine learning, robotics, and advanced engineering. This gap prevents individuals, especially those from marginalised communities, from acquiring the necessary knowledge and expertise to pursue careers in these rapidly advancing fields, limiting their opportunities for economic advancement and contribution to technological innovation. Unless there is a rapid and intentional intervention, the gap will continue to increase, putting young African women at a further disadvantage compared to their counterparts in other parts of the world.

This gap in the academic and advanced research fields progressively widens as women transition from undergraduate to postgraduate studies. Studies show that women comprise only 24% of full-time professors in the United States and 29% in the United Kingdom. Furthermore, research indicates that women are less likely to be promoted to senior faculty positions due to factors such as gender bias, a lack of mentorship opportunities, and work-life balance challenges. This underrepresentation also extends to management positions in educational institutions, with only 30% of university presidents being women. While these statistics are from advanced countries, they provide insight into some of the challenges STEM women experience, particularly in Africa.

Research Funding

Female researchers often face systemic barriers in securing funding for their projects, which affects their contribution to STEM projects and initiatives and the advancement of their careers. It has been established that female applicants receive less research time than males, and when they get grants, it is usually in small amounts. This makes their research papers shallow and hampers their ability to contribute significantly to their respective fields.

Global Competitiveness

Despite visible advancements, African women in STEM fields still need to be globally competitive through more advanced technologies, state-of-the-art laboratories, and engagement in cross-border research. Engaging in worldwide science and technology development activities without a level playing field becomes a challenge. Closing this gap is crucial for creating a more inclusive and competitive global STEM community.

Sustainability of Initiatives

Many initiatives targeting women in STEM are often developed as short-term, project-based programmes. While these programmes may succeed initially, they usually need help to maintain continuity and impact once the initial funding is exhausted. This support gap can slow down long-term effectiveness and sustainability, creating a cycle of starting and stopping meaningful and realistic long-term outcomes. This challenge affects the individuals involved and the overall progress in increasing diversity and representation in STEM fields.

These challenges disproportionately affect young girls in rural and underserved communities. They often encounter additional barriers such as cultural norms, early marriage, and household responsibilities, further limiting their access to education. Consequently, these obstacles lead to higher student dropouts among girls, perpetuating the cycle of limited educational opportunities and hindering their potential for success.

OVERCOMING BARRIERS

In the past few decades, African women have made significant strides in STEM, especially in education. Initiatives such as scholarships, mentorship programmes, and community-based efforts have contributed to the remarkable progress in encouraging more women to pursue STEM careers. As a result, there has been a noticeable increase in the enrolment of girls in STEM-related courses.

Despite this progress, significant barriers still impede women's full participation in STEM. Cultural norms and gender stereotypes continue to dissuade women from pursuing STEM careers, and the lack of female role models, particularly in leadership positions within STEM fields, further restricts young women's aspirations.

To address these challenges, educational systems must adapt and provide more supportive environments for women in STEM. For example, review for more inclusive curricula with practical experience and leadership development opportunities.

Transitioning from education to leadership roles in STEM remains challenging for many African women due to workplace dynamics that often reflect societal gender biases. Addressing these biases through policy changes and shifts in organisational culture is essential for progress in this area.

Moreover, targeted strategies such as continuous professional development, fostering supportive networks, and promoting women into visible leadership roles are crucial in ensuring that the progress made in education translates into success in the professional realm. These measures will contribute to a more inclusive and diverse STEM landscape, benefitting society.

MY STORY

My journey from the classroom to the boardroom in STEM fields has been challenging yet rewarding. As a black African woman, I have encountered various obstacles and triumphs that have shaped my professional trajectory. From navigating implicit biases to celebrating successes, my story demonstrates the resilience and determination required to succeed in a male-dominated industry. I hope sharing my experiences will provide valuable insight and inspiration for aspiring female leaders in STEM.

A Father's Ambition

Living in Nigeria during my formative years and unknown to me then, my secondary education at Methodist Girls' High School in Lagos laid the foundation for a STEM career. The school's rigorous academic schedule pushed me to excel in various subjects. However, my civil engineer father was pivotal in sparking and nurturing my STEM abilities. He dedicated his free time to teaching me and my friends mathematics, instilling in us a deep appreciation for problem-solving and analytical thinking.

My father, like many parents at the time, had envisioned a specific career path for me without my input. He determined I should become a chartered accountant and study accounting as my first degree. In contrast, I was equally determined to carve my own path, even though I had no idea where it might lead. I believed that pursuing accounting as a degree would be too restrictive, leaving little room for other enjoyable aspects of life. As a result, I chose to study psychology, a decision that would later prove beneficial in my transition to information systems.

From Psychology to Information Systems Engineering

Life took an unexpected turn when I pursued a postgraduate degree in the United Kingdom. Due to financial constraints, I found supporting myself through a three-year postgraduate degree in clinical psychology was challenging. As a result, I made the practical choice to pursue a more cost-effective one-year postgraduate degree in information systems engineering. This decision ultimately altered the trajectory of my career path.

Unlike some of my friends who had clear career paths from an early age, I needed a defined plan. It wasn't until later that I uncovered a deep passion for technology. This realisation led me to explore various roles in different industries, allowing me to weave together a diverse and rich tapestry of experiences in my career journey.

Navigating New Territories

I began my professional career as a desktop technician for a multinational bank, where I honed my technical skills and gained valuable experience in troubleshooting and IT support. This role served as the foundation for my career journey, leading me to transition into business analysis, project management, operations, and strategy, where I demonstrated my ability to adapt, change, and grow.

My tenure at renowned corporations such as Merrill Lynch, BP Oil, and the London International Financial Futures and Options Exchange (LIFFE) gave me invaluable exposure to cutting-edge global business practices and digital working tools. As a young African immigrant woman navigating the professional landscape,

I encountered and overcame significant cultural differences through dedication and perseverance.

Recognising the importance of effective communication, I immersed myself in learning the nuances of networking and small talk, skills at which the English excels. I proactively joined Toastmasters, a renowned leadership and communications organisation, to bridge the communication gap. Through my involvement, I acquired essential public speaking skills, which I have since imparted to many of my teams through mentorship, fostering their professional growth and development.

The Power of Adaptability

I encountered specific barriers throughout my career due to the absence of structured mentoring programmes. However, during the early stages of my career, I proactively assimilated the British way of doing things, which allowed me to settle in quickly. I owe a debt of gratitude to friends who supported me and facilitated my rapid adoption of certain facets of the British lifestyle. This experience helped me integrate into the new environment and significantly influenced my interests and thinking.

My approach to cultural immersion and my unwavering commitment to lifelong learning played pivotal roles in overcoming the challenges of being a black woman in the STEM field and an African in a foreign country. My adaptability and flexibility have helped me navigate past and present challenges and roles, underscoring the immense value of adaptability within the dynamic and ever-evolving STEM field.

Bringing Global Experience Home

After spending 20 years in the United Kingdom, I boldly returned to Nigeria, a move that I consider pivotal in my journey towards leadership growth. Joining an international bank in Nigeria exposed me to many new opportunities and challenges, providing me with invaluable insights into the intricacies of organisational culture and the complexities of internal politics within a company and their profound impact on my personal and professional growth as a leader.

The transition back to Nigeria could have been more seamless. Upon our return, many of us were initially met with admiration from our colleagues. However, this sentiment quickly turned to envy and resentment as colleagues perceived us as threats to their career advancement. This experience was a crash course in corporate acculturation and effective communication in leadership, and it has greatly enriched my understanding of navigating such delicate dynamics within an organisation.

Mentoring the Next Generation

Throughout my 30-year career journey in the UK and Africa, my most significant contribution has been my ability to mentor and inspire the next generation (my teams). Many are now in leadership positions, achieving great things they never envisaged at the time.

Having started as a junior graphic designer, I gradually expanded my expertise to include operations, business analysis, and project management. This diverse experience has shaped my career and leadership ability, enabling me to bridge the gap between technical and business integration.

My experience attests that resilience, adaptability, and continuous evolution are essential for pivoting across industries and fostering growth. This captures the essence of the African woman trailblazers in the journey of technological advancement and economic development within the region.

FROM CLASSROOM TO BOARDROOM – KEY SUCCESS FACTORS

As I reflect on our collective journey, I feel duty-bound to share the following advice with young African women who aspire to pursue a career in STEM and aim for leadership and board-level positions:

1. Identify a mentor with experience and insights relevant to your career aspirations.
2. Seek suitable sponsors within your organisation who are willing to advocate for your professional advancement.
3. Explore roles other than technical roles. Spread your tentacles to different business areas and acquire business skills.
4. Invest in lifelong learning through continuous professional development opportunities.
5. Develop political and cultural acumen to navigate diverse work environments effectively.

CONCLUSION

In summary, having listened to the stories of STEM women in board positions and their journeys in combination with research data and reports in the public domain, we can extrapolate the following: introduction to STEM at an early age will more likely lead to careers in STEM. Therefore, it is essential to create an enabling environment through the right government policies and develop frameworks that support positive learning techniques aligned with the way girls prefer to learn. In addition, encouraging and driving mentorship programmes with the help of the private sector and Non-governmental organisation (NGO) partnerships and the support of existing STEM woman leaders could pivot the trajectory of some of the new female entries into STEM. Finally, diversity, equity, and inclusion should become a reality not just on paper but also in corporate hiring, deployment, and promotion.

APPENDICES

Survey Analysis: African Women Trailblazers in STEM

This survey was conducted with 20 women board leaders to understand the experiences, challenges, and perspectives of African women who have achieved leadership status in STEM. The aim was to gather insights about their journey, critical success factors that catapulted them into leadership positions, and their advice for the next generation of women STEM leaders.

SURVEY QUESTIONS AND RESPONSES

1. *At what stage of your education or career did you enter the STEM field?*

80% entered the STEM field in high school, 10% entered during university, and 10% made mid-career transitions into STEM.

2. *Which of the following was the most significant challenge in your STEM career progression?*

27% cited a lack of mentorship, 20% for gender bias, 20% for limited networking opportunities, 20% for structural/organisational barriers and 13% for cultural expectations.

3. *How important was mentorship in your journey to a board position? (1 on a scale of 1–5)*

47% rated mentorship as very important, 20% rated it as moderately important, 20% rated it as less important. The Average rating was 2.27.

4. *Have you benefited from corporate policies specifically aimed at advancing women in STEM?*

40% reported no benefit from such policies, 27% benefited somewhat, 20% benefited significantly, 13% reported such policies didn't exist in their workplace.

5. *In your opinion, how has the landscape for African women in STEM leadership changed over the past decade?*

60% believe it has somewhat improved, 20% believe it has significantly improved, 13% are unsure.

Please share a pivotal moment or experience that significantly impacted your career in STEM.

The overall themes centre around these pivotal perspectives:

1. Taking on management or leadership roles.
2. Cross-functional experiences.
3. International work experiences.

4. Strong family support.
5. Formal mentorship programmes.

Based on your experience, what advice would you give to young African women aspiring to leadership roles in STEM, and what changes do you believe are needed in the industry to support their advancement?

Key Advice For Young Aspiring Females in STEM:

a) Seek mentorship and sponsorship early in your career.
b) Build strong networks. Learn about networking early in your career.
c) Continuously learn and develop expertise in your areas and acquire general business skills.
d) Be resilient and embrace challenges.
e) Take initiative and "raise your hand" for opportunities.
f) Stay true to yourself and your goals.

Suggested Changes For Organisations, Educators and Regulatory Bodies:

a) Implement diversity-promoting policies.
b) Improve STEM education curriculum.
c) Create more structured mentorship programmes.
d) Develop family-friendly workplace practices.
e) Increase the visibility of women in STEM leadership roles.

ANALYSIS OF THE SURVEY

The survey reveals that though progress has been made, gaps remain to be plugged, and significant challenges remain for African women in STEM. From the participants' stories and my story, it is clear that early exposure to STEM fields, mentorship, and supportive corporate policies are key factors in career advancement. Given the diverse challenges, there is a need for a consolidated and comprehensive approach to support women in STEM, including addressing cultural biases, improving networking opportunities, and breaking down structural barriers. That said, the positive perception within the landscape is encouraging, though varied experiences with corporate policies suggest that more effective solutions are needed. Many women strongly emphasised proactive career management, continuous learning, and resilience, which indicates the need to step out of one's comfort zone.

To foster the next generation of African women leaders in STEM, focus should be placed on early STEM education, creating robust mentorship programmes, implementing effective corporate policies, and increasing the visibility of women in leadership roles. These efforts and addressing cultural and structural barriers can create a more supportive environment for women to thrive in STEM fields across Africa.

REFERENCES

African Union. (n.d.). Gender Equality and Development | African Union. Retrieved from African Union: https://au.int/en/gender-equality-development

Editorial. (2022, February 22). Female scientists in Africa are changing the face of their continent. *Nature*, 602. Retrieved from Nature: https://www.nature.com/articles/d41586-022-00492-x

Rose-Innes, A. (2022, August 3). Africa has highest proportion of Female STEM graduates. Retrieved from Women In Science: https://www.womeninscience.africa/africa-has-highest-proportion-of-female-stem-graduates/

UNESCO. (n.d.). Girls and Women's education in Science, Technology, Engineering and Mathematics (STEM). Retrieved from UNESCO.org: https://www.unesco.org/en/gender-equality/education/stem

4 Cybersecurity Risk Governance in Africa

An African Female Director's Perspective

Eneni Oduwole

INTRODUCTION

Role of the Board in Cybersecurity Governance

Cybersecurity has become a global concern for organizations in an increasingly digital world. Africa, undergoing rapid digital transformation, faces unique cybersecurity challenges requiring robust governance frameworks. Including women directors on boards, particularly in cybersecurity, is not just a critical aspect of this governance but an urgent need. African women directors bring diverse perspectives, enhance decision-making processes, and contribute to a more inclusive and practical approach to cybersecurity. This chapter underscores the importance of African women directors' involvement in cybersecurity governance, highlighting their potential impact on organizational resilience and security.

As cyber threats grow in sophistication and frequency, board members must ensure robust cybersecurity measures are in place to protect the organization's assets and reputation. This involves understanding the evolving threat landscape, overseeing the implementation of effective cybersecurity policies, and ensuring compliance with regulatory requirements.

Concepts of Cybersecurity Governance

Organizations must develop and implement robust cybersecurity governance frameworks to manage cybersecurity risks effectively. These frameworks should clearly define the concepts explained herein.

- **Risk Management**

The process for continuously identifying, assessing, and prioritizing cybersecurity risks must be clearly defined. This is done by identifying prevalent and likely cyber risks that could threaten an organization's cybersecurity, assessing the likelihood of occurrence and impact on the organization, whether financial or non-financial,

DOI: 10.1201/9781003540618-4

such as reputational damage or regulatory sanctions, and prioritizing identified cyber risks for proper mitigation. The board's prerogative is to decide and approve the appropriate risk treatment for each identified risk, whether to treat, transfer, share, or accept the risk-based on its risk appetite.

- **Board and Executive Oversight**

The board is responsible for building a solid cybersecurity culture in any organization. In line with global best practices, the board and executive management are accountable for cybersecurity governance and oversight of the organization's cybersecurity strategy. Therefore, the board must prioritize cybersecurity organization-wide by embedding it into its culture and decision-making mechanisms.

- **Policy and Frameworks**

Cybersecurity governance is well-implemented when appropriate policies, standard operating procedures, and other documentary guidelines exist. These documents define the expected and acceptable behaviours, protocols, and security standards for managing cybersecurity risks. Globally accepted frameworks that can be referenced or used as guides for developing relevant policies, procedures, and protocols include the NIST Cybersecurity Framework or ISO/IEC 27001 standard, which provide structured approaches for cybersecurity risk management and governance.

- **Compliance and Legal Requirements**

The board must ensure its organization complies with relevant local and international data protection and cybersecurity laws such as the GDPR, POPIA, and CCPA. Other industry-specific cybersecurity standards, such as the PCI-DSS for payment security, should also be leveraged.

- **Incident Response and Resilience**

Cybersecurity governance requires the creation of robust incident response plans detailing the organization's response to and recovery from cybersecurity incidents such as data breaches or ransomware attacks. It should also emphasize building organizational resilience, ensuring effective management of any cyberattack, and timely resumption of business with minimal or no downtime or loss of data.

- **Continuous Monitoring and Reporting**

Continuous monitoring of cybersecurity performance and incidents is crucial. Therefore, it requires regular reporting of critical cyber risk metrics and threats to the board and senior management. It is also very important that the board ensures that independent audits and assessments are carried out periodically to ensure the cybersecurity governance framework is effectively implemented and aligned with organizational objectives.

- **Stakeholder Engagement**

The board must ensure that all internal and external stakeholders relevant to its cybersecurity governance, such as IT teams, business units, external partners, and regulators, are appropriately engaged to ensure a comprehensive risk management approach to cybersecurity. It must also ensure that vendor or third-party risks are well managed and that external vendors and partners comply with the organization's cybersecurity policies. Third-party risks can expose the organization to cyber threats.

- **Training and Awareness**

For a solid cybersecurity culture to be nurtured and maintained organization-wide, the board should ensure its cybersecurity framework and approach mandate a culture of cybersecurity awareness across all levels through periodic, regular training to ensure employees understand their role in maintaining the organization's cybersecurity strategy. It is also crucial that board members are educated on emerging cybersecurity threats and best practices to make informed decisions.

- **Cyber Risk Appetite and Strategy**

An organization's risk appetite defines the size of the risk it is willing to take at any given time as it relates to any transaction or business activity. Therefore, the board must define the organization's cyber risk appetite. It must also ensure that the organization's cybersecurity framework and strategy align with its broader business objectives, ensuring that risk management efforts support its overall mission and vision.

- **Data Governance and Privacy**

Protecting sensitive data, such as customer information and intellectual property, is critical to cybersecurity governance. Organizations must implement controls to safeguard data from unauthorized access and ensure compliance with data privacy laws and regulations, balancing the need for security with individuals' rights. The board has a crucial role in enforcing this in an organization.

- **Threat Intelligence and Adaptation**

A robust cybersecurity governance framework must stipulate the use of threat intelligence to proactively predict and pre-empt cyber threats, so that appropriate treatment of such risk exposures is implemented before they materialize. The board must ensure its governance frameworks are flexible and continuously updated to adapt to the rapidly changing cybersecurity landscape, including new threats, technologies, and regulatory requirements.

- **Ethics and Corporate Social Responsibility**

The board must ensure its cybersecurity practices are ethical and align with globally accepted ethical standards. It must also promote trust with customers, employees, and other stakeholders. The board must emphasize the broader impact of its cybersecurity practices on society, including how data is handled, how privacy is respected, and how it contributes to the overall security ecosystem.

A Board's Strategic Approach to Cybersecurity Governance

Every board is responsible for ensuring that its organization's exposure to cybersecurity risks is assessed and managed effectively. The board must collaborate with relevant stakeholders, especially IT leaders, to understand its most significant cybersecurity risks. Therefore, continuous monitoring and regular reporting of the organization's cybersecurity risk profile highlight emerging threats and the organization's defences. The board should also develop and implement a well-defined and tested cyber incident response plan. This onerous responsibility highlights the importance of educating board members on cybersecurity trends and threats.

By prioritizing cybersecurity risk governance, African women board members can mitigate potential damages, safeguard stakeholder interests, and promote a culture of security within the organization. Effective oversight and strategic guidance are essential to navigating the complexities of cybersecurity threats.

The Perspective of an African Female Director

Focus on the impact that African female directors can make in achieving effective cybersecurity governance is necessitated by the need to break and annihilate some biases that are common in African societies. One of these biases is the perception that women lack the technical expertise to excel in STEM careers. As cybersecurity is globally regarded as a highly technical field, it is not surprising that the bias exists that women, including African female board members, may not possess the technical skills or knowledge needed to govern cybersecurity effectively. This mindset does not consider that cybersecurity governance is about risk management, strategy, and policy areas in which African women have excelled.

Furthermore, African female board members who oversee cybersecurity governance have to disprove the global stereotype that men are most suited for technical roles and responsibilities such as this, especially in Africa, where gender roles influence the trust and perception of women in leadership roles. As a result, a typical African female board member responsible for cybersecurity governance may face the dual challenge of gender and racial biases and perceptions that she is less competent or authoritative in cybersecurity matters than her male counterparts.

African women are underrepresented on most boards, so they may not be fully recognized for their leadership role in cybersecurity governance. This can impact their ability to influence decisions in this critical area. However, as more African women break these barriers and the importance of diversity in governance becomes increasingly evident, such biases may disappear.

LITERATURE REVIEW

The board's involvement in cybersecurity governance is vital for ensuring comprehensive risk management and organizational resilience. According to Abrahams and Fitzgerald (2019), board oversight in cybersecurity is not merely a technical necessity but a strategic imperative. Their research indicates that proactive board engagement can significantly enhance an organization's agility and ability to respond to and recover from cyber incidents. Similarly, Williams and Mugisha (2020) assert that board-level commitment to cybersecurity fosters a culture of security that permeates the entire organization.

African boards encounter specific challenges in cybersecurity governance, often rooted in limited resources and expertise. Afolayan and Akinsola (2021) highlight the scarcity of cybersecurity professionals in Africa, complicating efforts to integrate cybersecurity into boardroom discussions effectively. This expertise gap is further exacerbated by the rapid evolution of cyber threats, requiring continuous learning and adaptation. In addition to expertise, financial constraints pose a significant hurdle. As noted by Nkomo and Sibanda (2020), many African organizations operate on tight budgets, which limits their capacity to invest in advanced cybersecurity measures and training for board members. This financial limitation often leads to reactive rather than proactive cybersecurity strategies, increasing the risk of significant breaches.

The regulatory environment in Africa regarding cybersecurity is evolving, with varying degrees of maturity across different countries. Nguyen and Ntim (2020) discuss the regulatory disparities within the continent, noting that while some countries have established robust cybersecurity frameworks, others are still in the nascent stages of development. This uneven regulatory landscape challenges boards striving to maintain compliance and adopt best practices across multinational operations.

Despite these challenges, regulatory developments are driving increased board attention to cybersecurity. Olayinka and Temitope (2022) emphasize that recent regulations in countries like Nigeria and South Africa are compelling boards to adopt more stringent cybersecurity measures. These regulations often mandate regular risk assessments, incident reporting, and the implementation of security controls, thereby embedding cybersecurity into the governance framework.

The literature proposes several strategies to address the challenges and enhance cybersecurity governance. Continuous education and training for board members emerge as a recurring theme. Afolayan and Akinsola (2021) advocate for mandatory cybersecurity training programmes tailored to board members' needs, ensuring they can make informed decisions and provide effective oversight.

Another critical strategy is integrating cybersecurity into the overall corporate strategy. Abrahams and Fitzgerald (2019) argue that cybersecurity should not be siloed as a purely technical issue but should be woven into the strategic fabric of the organization. This integration ensures that cybersecurity considerations are factored into business decisions, reducing risks and enhancing resilience.

The establishment of dedicated board committees for cybersecurity is also recommended. According to Nkomo and Sibanda (2020), having a specialized committee allows for focused attention on cybersecurity issues and fosters more profound expertise among its members. This committee can oversee the development and implementation of cybersecurity policies, conduct regular risk assessments, and ensure compliance with regulatory requirements.

CHALLENGES OF AFRICAN FEMALE DIRECTORS IN CYBERSECURITY GOVERNANCE

Skills Gap and Access to Expertise

There is a global shortage of cybersecurity professionals and experts, which is more pronounced in Africa. The significant disparity in the skills gap and access to expertise heightens when considering the number of African women in this field. According to the (ISC)² Cybersecurity* Workforce Study:

> *Globally, women represented about 24% of the cybersecurity workforce in 2019. In regions such as Africa, where STEM and cybersecurity gender gaps are more pronounced, estimates are significantly lower, typically cited between 7% and 11% in various reports and expert discussions.*

This skills gap and lack of expertise in the African region impact the number of females who can become qualified professionals in this field, as cultural biases focus on spending more money on training the boy child. As a result, African female directors may find fewer tailored opportunities to gain advanced cybersecurity skills, particularly in governance, risk management, and compliance (GRC) specific to their roles as executives or board members.

Limited or Underdeveloped Cybersecurity Infrastructure

Many African countries have prevalent issues with outdated or limited digital infrastructure, making enforcing robust cybersecurity governance frameworks more challenging. This is particularly challenging for female directors on the continent, who often encounter biased or prejudiced notions about their leadership abilities, particularly in male-dominated professions such as cybersecurity. They sometimes experience doubts from their male counterparts about their competence and authority, thus undermining their ability to serve effectively as leaders in cybersecurity

* (ISC)² is an international nonprofit membership association for cybersecurity professionals.

governance. The work of Moraka, N. (2023) posits that African women in executive roles face challenges related to gender stereotyping and implicit bias in training and development, which diminishes their authority in leadership spaces traditionally dominated by men.

This further constrains their awareness and knowledge of the evolving threat landscape, which is riddled with sophisticated forms of ransomware, phishing, and identity theft, as well as innovative cybersecurity tools and solutions, becoming a handicap to effective cybersecurity governance. Many African organizations involved in international trade or services experience unprecedented cyber threats from countries outside the continent. Any director's skill gap in managing such complexity, primarily because of the different cybersecurity regulatory frameworks and technological infrastructures, will be disadvantageous for the organization. In Africa, this unfavourable situation is usually considered worse if such a director is female.

Regulatory Complexity and Data Protection Requirements

Africa has various cybersecurity regulations, with each country having different data protection, privacy, and cybersecurity guidelines. Female directors need to have in-depth knowledge and understanding of the various regulatory environments in which their organizations operate and how they interact with international standards like the GDPR.

Once again, the skill set and expertise required to oversee the implementation and enforcement of cybersecurity policies in a context where laws are either new or underdeveloped present additional challenges for an African female director.

Cultural and Gender Bias and Leadership Gaps

Systemic gender biases in leadership roles within the tech industry exist in Africa, and this is prevalent in the cybersecurity profession. Female directors sometimes face scepticism about their technical capabilities and have fewer role models or mentors in cybersecurity governance, especially as cybersecurity and governance board committees are usually male-dominated worldwide. Female directors, particularly African women, may face challenges asserting influence or being taken seriously in cybersecurity discussions.

Funding and Resource Constraints

Many African organizations have limited budgets for cybersecurity initiatives, and female directors on the continent may struggle to secure the necessary investment for robust cybersecurity governance. This impacts their effectiveness because prioritizing cybersecurity governance in organizations where other pressing business needs, such as operations, marketing, and planning, take precedence can be technically challenging.

Adoption of Emerging Technologies

With the rise of technologies such as AI, blockchain, and IoT, female directors must understand the cybersecurity implications of these technologies. Adopting these innovations without a solid technical foundation in cybersecurity governance can expose organizations to new risks. The disparity with the varying levels of digital transformation across organizations within the continent complicates governance for directors overseeing cybersecurity because of the different levels of cybersecurity maturity.

AFRICAN WOMEN DIRECTORS IN CYBERSECURITY RISK GOVERNANCE

Diversity and Inclusivity

One key reason for involving African women directors in cybersecurity governance is to enhance diversity and inclusivity. Research has consistently shown that diverse boards make better decisions. According to Nielsen and Huse (2010), gender diversity on boards is associated with improved monitoring and strategic decision-making. Women bring different experiences and perspectives to the table, which can lead to more comprehensive discussions and innovative solutions to complex problems such as cybersecurity.

In Africa, where cultural and social dynamics often influence business practices, women directors can provide unique insights into ethical behaviour, consumer behaviour, regulatory environments, and social expectations. This diversity of thought is crucial in understanding and mitigating cybersecurity risks, which are multifaceted and constantly evolving. By fostering a culture of inclusivity, organizations can better anticipate and respond to cyber threats, ultimately enhancing their resilience.

Addressing Cybersecurity Skills Gap

The cybersecurity skills gap is a significant challenge in Africa, with a shortage of trained professionals to address the increasing cyber threats. Women directors can play a vital role in bridging this gap. By advocating for and supporting initiatives that promote cybersecurity education and training, women directors can help build a pipeline of skilled professionals. This is particularly important on a continent where educational and professional opportunities for women are often limited.

Programmes that encourage young women to pursue careers in STEM (Science, Technology, Engineering, and Mathematics) and cybersecurity can have a long-term positive impact. Women directors, serving as role models and mentors, can inspire and guide the next generation of female cybersecurity professionals. This helps address the skills shortage and promotes gender equality in a traditionally male-dominated field.

Promoting a Culture of Security

Board members play a crucial role in shaping the organizational culture. Women directors can be instrumental in promoting a culture of security within their organizations. Westerman et al. (2020) state that organizational culture is crucial to cybersecurity effectiveness. A culture that prioritizes cybersecurity encourages employees at all levels to take responsibility for security measures and practices.

Women directors can champion policies and practices integrating cybersecurity into the organization's core values. This includes advocating for regular training and awareness programmes, ensuring cybersecurity is a standing item on board agendas, and promoting transparency and accountability in reporting and responding to cyber incidents. African women directors can significantly enhance their organizations' cybersecurity by fostering a proactive and security-conscious culture.

Enhancing Risk Management and Governance

Effective cybersecurity governance requires robust risk management frameworks. Women directors bring valuable skills and perspectives to risk management processes. According to the *Harvard Business Review* (2019), women tend to be more risk-aware and cautious in their decision-making, which can lead to more thorough assessments of potential threats and vulnerabilities.

Involving women directors in cybersecurity governance ensures that risk management strategies are comprehensive and balanced. Their attention to detail and propensity for collaborative decision-making can lead to more effective identification, assessment, and mitigation of cybersecurity risks. This is particularly important in Africa, where the threat landscape constantly evolves, and organizations must be agile and adaptive in their risk management approaches.

Building Trust and Stakeholder Confidence

Trust and confidence are critical components of effective cybersecurity governance. Stakeholders, including customers, investors, and regulators, must be confident that organizations are responding appropriately to protect their data and systems. Women directors can play a crucial role in building and maintaining this trust.

Research by Adams and Ferreira (2009) indicates that women on boards are more likely to engage in behaviours that enhance transparency and accountability. Their involvement in cybersecurity governance can ensure that organizations adopt best practices, comply with regulatory requirements, and communicate effectively with stakeholders about their cybersecurity efforts. This transparency and accountability are essential in building trust and confidence, which are vital for organizational success in a digital economy.

Driving Innovation and Strategic Initiatives

Cybersecurity is not just about defence but also about innovation and staying ahead of potential threats. Women directors can drive innovation by encouraging the adoption of new technologies and strategic initiatives that enhance cybersecurity. Their diverse perspectives and collaborative leadership styles can lead to creative solutions and initiatives that address emerging cyber threats.

In Africa, where technological advancements rapidly transform business practices, women directors can advocate integrating innovative technologies such as artificial intelligence, machine learning, and blockchain into cybersecurity strategies. By staying ahead of the curve and embracing innovation, organizations can better protect themselves against evolving cyber threats and capitalize on new opportunities in the digital landscape.

THE ROLE OF CULTURE AND LEADERSHIP

Organizational culture and leadership dynamics play a crucial role in cybersecurity governance. According to Abrahams and Fitzgerald (2019), boards must cultivate a security-focused culture where employees at all levels understand their role in protecting the organization's digital assets. This cultural shift requires strong board leadership that emphasizes the importance of cybersecurity through consistent messaging and actions.

Leadership commitment is further reinforced by the involvement of the Chief Information Security Officer (CISO) or equivalent in board discussions. Afolayan and Akinsola (2021) argue that direct communication between the board and cybersecurity executives ensures that the board is well-informed about the organization's cyber risk landscape and the effectiveness of its cybersecurity measures.

AFRICAN SUCCESS STORIES

Case studies from various African organizations provide practical insights into effective board involvement in cybersecurity governance. Williams and Mugisha (2020) present a case study of a leading telecommunications company in Kenya that successfully integrated cybersecurity into its corporate governance framework. The company's board established a cybersecurity committee, invested in regular training for its members, and adopted a risk-based approach to cybersecurity. As a result, the company significantly improved its cybersecurity posture and resilience against cyber threats.

Another case study by Olayinka and Temitope (2022) examines a Nigerian financial institution that faced multiple cyberattacks. The board's proactive measures, including recruiting cybersecurity experts and implementing advanced security technologies, were pivotal in mitigating the impact of these attacks. These cases underscore the importance of board commitment and investment in building robust cybersecurity defences.

Several African women directors have made significant contributions to cybersecurity governance. One notable example is Funke Opeke, CEO of MainOne, a leading provider of innovative telecom services and network solutions in West Africa.

FUNKE OPEKE

Under her leadership, MainOne has invested heavily in cybersecurity infrastructure and practices, protecting its critical assets and customer data. Opeke's strategic vision and commitment to cybersecurity positioned MainOne as a trusted and reliable regional partner because of its emphasis on the importance of data governance and protection. The company has spearheaded initiatives to secure the region's data centres and internet infrastructure, working with regional cybersecurity frameworks and global standards.

Opeke has also advocated for Africa's technology self-sufficiency, including data protection and cybersecurity. Her efforts reflect the growing recognition of African women in cybersecurity governance in the continent's digital infrastructure. Funke Opeke's work in cybersecurity governance shows how African women in technology leadership positions can influence cybersecurity governance indirectly by securing infrastructure and advocating for strong governance policies in sectors that support cybersecurity.

LINDIWE MATLALI

Lindiwe Matlali is a social entrepreneur and the founder of Africa Teen Geeks, a non-profit organization that introduces African youth to coding, cybersecurity, and STEM education. Matlali focuses on the cybersecurity governance pipeline by ensuring African girls and women have access to cybersecurity education and careers. This initiative ensures that African women have the required skill sets to take on leadership roles in cybersecurity governance.

Her work focuses on creating a pipeline of future cybersecurity professionals, particularly women. This is very important because governance frameworks are strengthened by a diverse, skilled workforce that reflects the population it serves through Africa Teen Geeks, she equips young African girls and women with the knowledge necessary to take on leadership roles in cybersecurity governance in Africa.

To achieve her goals, she works closely with private companies and government agencies to advocate for policies that promote the inclusion of underrepresented groups in tech and cybersecurity. Her work demonstrates that governance extends beyond corporate boards and government agencies—it starts with education. Her efforts to develop future African female cybersecurity professionals help address gaps in representation.

YVONNE MUTHONI

Yvonne Muthoni is the Director of Cyber Security Services at Serianu, a leading cybersecurity firm in Kenya. Muthoni has been pivotal in driving cybersecurity awareness and solutions across East Africa.

Muthoni is directly involved in assisting organizations across East Africa to implement cybersecurity governance structures. She advises companies on regulatory compliance, risk management, and cyber incident response and ensures they adhere to best practices in cybersecurity governance. She is renowned for her advocacy for women in cybersecurity. She actively promotes the inclusion of African women in cybersecurity, emphasizing that they bring unique perspectives and leadership qualities to governance roles.

Yvonne Muthoni's leadership within a cybersecurity firm shows how African women shape corporate governance in cybersecurity. Her advocacy for women's representation is vital to creating more inclusive and effective governance structures.

DR OMOBOLA JOHNSON

Dr Omobola Johnson is an accomplished technocrat and Nigeria's former Minister of Communication Technology. During her tenure, she implemented Nigeria's first national ICT policy and played a significant role in cybersecurity initiatives, emphasizing governance, strategy, and the regulatory environment. She serves on multiple boards, including as Chairperson of the Alliance for Affordable Internet (A4AI).

As a minister, Dr Johnson led the development and implementation of Nigeria's cybersecurity policies. Her leadership was instrumental in developing the National Cybersecurity Policy and Strategy (NCPS), which laid the foundation for Nigeria's approach to cybersecurity governance. She advocates for cybersecurity governance in the private sector, serving on various boards and advising on its importance in corporate governance, particularly for tech companies and financial institutions.

Dr Omobola Johnson exemplifies how African women in government can directly influence national cybersecurity strategies and governance frameworks. Her ability to translate policy into actionable frameworks for the public and private sectors displays women's vital role in shaping Africa's cybersecurity landscape.

EDITH MWALE

Another example is Edith Mwale, a member of the Zambia Information and Communications Technology Authority (ZICTA) board. Mwale advocated for stronger cybersecurity regulations and policies in Zambia. Her efforts have led to the implementation of comprehensive cybersecurity frameworks that enhance the country's resilience to cyber threats.

FUTURE DIRECTIONS AND RECOMMENDATIONS

The literature offers several recommendations to enhance the board's role in cybersecurity governance in Africa. First, there is a need for increased investment in cybersecurity training and education for board members. This investment should be seen not as a cost but as a critical component of organizational resilience. Moraka, N., 2023, recommends that female directors be strategically assigned to important committees, such as nominations, audit, finance, risk, and remuneration, to acquire the technical expertise required for influencing decision-making.

Second, regulatory harmonization across African countries can provide more explicit guidelines and standards for boards to follow. Nguyen and Ntim (2020) suggest that regional cooperation and developing Pan-African cybersecurity frameworks could help address regulatory disparities and promote best practices. Governments, international organizations, and private sectors must collaborate to provide better access to cybersecurity tools and training for female leaders.

Third, fostering partnerships between the private sector, government, and academia can help address the expertise gap. Initiatives that encourage knowledge sharing and collaboration can enhance the overall cybersecurity ecosystem and provide boards with access to the latest insights and technologies. Promoting gender diversity in cybersecurity will help to address cultural and systemic gender biases and foster an inclusive environment for women in cybersecurity governance.

CONCLUSION

The board's role in cybersecurity governance is crucial for the protection and resilience of African organizations. While African boards face significant challenges, including limited expertise, financial constraints, and evolving regulatory landscapes, the literature highlights several strategies and best practices to enhance their effectiveness. Continuous education, strategic integration of cybersecurity, dedicated board committees, and a strong culture of security are essential components of robust cybersecurity governance. By adopting these practices, African boards can better navigate the complexities of cybersecurity and safeguard their organizations in an increasingly digital world.

Therefore, the involvement of African women directors in cybersecurity governance is essential for enhancing organizational resilience and security. Their diverse perspectives, leadership skills, and commitment to promoting a security culture can significantly strengthen cybersecurity frameworks. Women directors can drive positive change and innovation in cybersecurity governance by addressing the cybersecurity skills gap, enhancing risk management, and building stakeholder confidence.

In a continent where digital transformation is rapidly evolving, the inclusion of women directors in cybersecurity governance is not just a matter of gender equality; it is a strategic imperative. Organizations that embrace this diversity and leverage the unique contributions of women directors are better positioned to navigate the complexities of cybersecurity and thrive in the digital age. As Africa continues to advance technologically, the role of women directors in cybersecurity governance will be increasingly important, ensuring a secure and prosperous future for organizations across the continent.

In conclusion, African female directors in cybersecurity governance must navigate a complex landscape that combines technical, regulatory, cultural, and gender-specific challenges. Overcoming these barriers requires targeted efforts in education, mentorship, and collaboration across sectors.

REFERENCES

Abrahams, R., & Fitzgerald, M. (2019). Cybersecurity governance: A board's perspective. *Journal of Business Ethics, 160*(2), 399–414.

Adams, R. B., & Ferreira, D. (2009). Women in the boardroom and their impact on governance and performance. *Journal of Financial Economics, 94*(2), 291–309.

Afolayan, T., & Akinsola, T. (2021). Bridging the cybersecurity skills gap in Africa: The role of corporate governance. *African Journal of Information Systems, 13*(3), 120–137.

Harvard Business Review. (2019). The risk management approaches of women and men: What works best? Retrieved from https://hbr.org/

Moraka, N. (2023). Breaking the glass ceiling: unveiling the talent management gap in boardrooms and its impact on low women representation. *International Journal of Research in Business and Social Science (2147- 4478), 12*, 119–133. https://doi.org/10.20525/ijrbs.v12i8.2895

Nguyen, P., & Ntim, C. G. (2020). Regulatory impacts on cybersecurity governance in Africa. *Journal of Cyber Policy, 5*(1), 56–74.

Nielsen, S., & Huse, M. (2010). The contribution of women on boards of directors: Going beyond the surface. *Corporate Governance: An International Review, 18*(2), 136–148.

Nkomo, S., & Sibanda, S. (2020). Financial constraints and cybersecurity investments in African firms. *African Journal of Management, 6*(4), 289–310.

Olayinka, A., & Temitope, O. (2022). Cybersecurity governance in Nigerian financial institutions: A case study. *Journal of Financial Regulation and Compliance, 30*(1), 43–59.

Westerman, G., Bonnet, D., & McAfee, A. (2020). *Leading digital: Turning technology into business transformation*. Harvard Business Review Press.

Williams, K., & Mugisha, J. (2020). Telecommunications and cybersecurity governance in Kenya: A case study. *Journal of African Business, 21*(2), 152–170.

5 A Woman in Health Leadership

My Story, Herstory

Eleanor Nwadinobi

A Woman in health leadership: My story, Herstory

Nigerian writer and poet, Chinua Achebe, in his book *Things Fall Apart*, said, "Until the Lions have their own historian, the history of the Hunt will always glorify the hunter." I will be roaring through my story, her story, a story that I own and tell for posterity.

I was born in the United Kingdom to a Nigerian father and a Jamaican mother. I understood from an early age the issues of mixed and diverse cultures. My father was one of Nigeria's pioneer veterinary surgeons, as documented in the Nigerian Government Gazette of 1972. It took a lot for his community to understand why he had gone abroad to study medicine for treating animals instead of humans.

My mother studied nursing in London and was part of a group of pioneer Jamaican nurses to settle in Nigeria. She was renowned in our community for introducing Jamaican cuisine, fresh fruit juices, and exotic fruits like ackee, breadfruit and June plum. I remember how amazing my mum looked in her uniform: the starched fascinator-style nurse's cap, elastic belt, institutional badges and glistening white shoes.

I was born not preferred by the Igbo culture because, at that time, preference for sons was prevalent. This meant that daughters were seen as less important than sons and received less education. Thankfully, my parents went against the grain of the Igbo culture and ensured that I received the best education possible as a girl.

My mother told me that my interest in medicine was sparked by empathy for a victim of a traffic accident when I was five. My mum narrated the scene to me where a car was involved in an accident, my parents went out to help, and when they returned, I immediately announced to them that I wanted to be a doctor. I was too young to remember that incident, but my parents found it prophetic.

My mother taught me to be confident, reassuring me always that I would not need anyone to validate me. My skills in public speaking were honed on account of poetry and drama outings in secondary school. Sr Dympna Drury, my school principal, deserves special mention in this regard. Through faith, hard work and sheer determination, I pursued a career in medicine and ended up being the first female doctor from my village. More importantly, more young girls have since been motivated to become doctors. I studied sciences at a time when, if you had very high marks in the sciences, you would naturally opt for medicine. Years later, at the age of 17, I found

DOI: 10.1201/9781003540618-5

myself at the University of Nigeria after taking an entrance exam and having my name published in the national newspaper. Those years were a beautiful experience, both for learning medicine and cultivating lifelong friendships, including with my best friend, who is now my husband of 42 years and counting.

Knowing that my parents had sacrificed so much to get me into medical school, I delayed acceptance of my husband's marriage proposal so that I could graduate with my maiden name and honour my parents. My parents had the joy of proudly displaying my graduation picture with my certificate bearing my maiden name. Six months later, I got married and took on my husband's name as a matter of choice.

I had a lot of challenges in the course of studying medicine. It was not the easiest of courses to study; it was lengthy and required a lot of discipline. I learnt discipline through the study of medicine. Everybody knows it is not easy, but it is about putting your head down. The practice of medicine is most challenging when you are desperately trying all you can to save a life, you do all you can humanly do, and still, the patient dies.

It has been quite an interesting journey. I started out in the field of medicine, within the speciality of anesthesia. I practised as an anaesthetist in Enugu and also in the UK, mostly in the south east Kent health authority in the UK. It wasn't easy being a resident doctor in the UK, especially in the field of anaesthesiology. The sharpening of the reflexes and the joy of saving lives made it all worthwhile. Carrying a crash bleep was, at first, a novel experience. Along with it came the excitement of leaping into resuscitation mode whenever it went off – bounding up the stairs two at a time or racing down corridors with my lab coat parachuting behind me. The practice as it was at that time, meant that I covered the provision of services in the theatre, in the obstetrics ward, especially for epidural injections, and in the intensive care unit. Additional services included calls out to the Kent Air Ambulance.

My children at that time were still quite young. In fact, I started my residency when my second child was only six weeks old. Thank God for mothers. My mum came to the rescue and took over the care of my infant daughter and toddler son. I have memories of bending down to intubate a patient and my breast milk overflowing like a tap. The patient was the priority, so the milk was left to dry on my shirt until the surgery was over and the patient was awake. Sometimes the theatre nurse would find an extra set of scrubs to throw over me. Colleagues coming on afternoon duty, I would see my three-year-old son quite literally taking his nana for a walk – he running ahead on the leash while my mum tried to keep up while hanging on to the pushchair.

All of this was combined with the challenges of bringing up my first two children. When they were old enough, I was relieved to find a lovely nursery school and daycare. Though the daycare fees were astronomical, even for me as a resident doctor, the convenience of a school bus which picked them up early in the morning and dropped them off in the evenings was lifesaving. Once back from work, I started preparing for dinner, getting ready for the next day's school and making sure that I had enough rest myself to face another day at work.

There were days when the children would have one ailment or another. My limited knowledge in other fields of medicine stretched the receptacles of my imagination. I have to confess that trusting another colleague to provide the best health care for my children was sometimes challenging. The symptoms would drive my imagination to the worst possible diagnosis, and it would take the calming influence of the attending colleague to inform me that it was a simple upper respiratory tract infection, and not some highfalutin, unpronounceable disease.

Looking back, I am grateful for the support system that my church provided beyond my immediate family. They made up the ever-present constellation that became my coping strategy. I looked forward to church family invitations for group excursions to theme parks, picnics and the zoo.

It wasn't until I relocated back to Nigeria that I could even think about having more children. The luxury of the extended family offering to babysit and the availability of affordable household staff made things easier. One could plan better for school runs and after-school clubs, and I could get back into a social life for both for myself and my husband.

I thank God that we had the gift of children while we were still young and had the energy to attempt the balancing act, mostly learning while on the job. I did not always get it right, however, despite well-intentioned plans.

Now that our children are grown and have flown the coop, I can look back on those years, marvelling at how my husband of over 40 years and I did it. When our granddaughters were just learning to make sentences, I taught them already to affirm to themselves, "I am beautiful, I am powerful, and I am strong," while tightening their little fists and displaying their biceps.

It is my prayer that our medical schools will include more formal training and courses on work-life balance and that work environments more conducive to achieving this will be provided for the next generation of medical practitioners. As they take over the baton and perform their roles on the "school of life" stage, their handling of work-life balance will hopefully be a huge improvement on ours.

Some events happen that are so momentous that they literally change the course of our lives.

When I came back to Nigeria, I started a partnership private practice called The Tabitha Infirmary (TTI) with a beloved sister Prof Ifeoma Okoye. She and I became founders of a women's fellowship, Blossoms of the Vine Fellowship. BVF remains today a place of spiritual nourishment for several young ladies.

As a doctor running the private hospital, I hired a security staff for the hospital gate who always called me nurse. His thinking was that women were nurses and men were doctors.

While I was working there, I discovered a situation that was prevalent among women in the south east which I was not previously aware of. That was harmful traditional practices meted out to widows. I began asking questions and also chose to do independent research with recommendations for a solution. This led me to go into human rights interventions.

My husband, my Okey, a devoted father, was my rock and number one cheerleader in supporting my desire to achieve my career aspirations. When I made the difficult decision to leave clinical medicine to pursue public health and development, he was most supportive. He understood my passion for women's rights. Though it was challenging for the family to depend on only one income while I pursued a master's degree in human rights, with his support, we did it.

When I stopped practising clinical medicine, my parents, who had spent their resources seeing me through medical school, were disappointed. However, when they saw the tremendous satisfaction I got from impacting not just individual patients one at a time, but impacting on whole populations, they understood my passion.

WIDOWS ACTIVISM – MY TURNING POINT

Aka – not her real – name was a friend through our church fellowship. She was around 20, a banker when her husband died in a car crash. As friends would do normally, a group of us drove down to her village in the south of Nigeria to pay our respects. When we got to her village, we saw somebody on the side of the road who we thought was mentally challenged. We all walked past this person and went to the main house to pay our condolences.

But the family pointed us to the women we passed, sitting on the ground..None of us had recognised our friend. She sat on the ground, covered in dust, disoriented. It was then we were told that this was the custom: from the time her husband died, she had not been allowed to bathe; she had to sit in the sun; she hadn't been allowed to comb her hair. I felt very uncomfortable. On the drive home, I asked my friends questions. They told me to stop talking about it; it was taboo to even discuss it. But I kept asking questions, and I had a "burden" to do something.

What Inspired My Switch from Clinical Medicine to Health and Human Rights?

My curiosity regarding harmful traditional practices towards widows led me to carry out research on the prevalent widowhood practices which existed in the four Southeastern States of Nigeria in 1994 – Abia, Anambra, Enugu and Imo. Today, there is an additional state called Ebonyi. I was motivated by a young man, Godwin Nwaoguwu, who had carried out research on another harmful practice in Igbo land called the Osu caste system, which is a caste system practice.

The results of the independent research were sobering as I discovered a lot of cultural practices which were dehumanising and clearly an infringement on the rights of women and girls. There was a general belief that women were responsible for the death of their husbands and, in some cases, accused of being witches. Therefore, very punitive mourning rites and practices were imposed on widows, supposedly to dissuade other women from killing their husbands. The burden I felt for widows gave me the courage to stand up and speak out regarding these practices, even though it was considered a taboo subject. More than just speaking up, I felt the need to look for ways to help them.

I summarised the independent findings that widows go through into three distinct issues when they lose their husbands, and I called them the 3D's. The first D is embodied as dethronement. The meaning of this is that a large premium is placed on marriage as lifting a girl from a low status and enthroning her to a place of status and recognition. However, the moment her husband dies, not minding the cause of death, she immediately loses this status and is "dethroned." The symbolic dethronement is practised by asking the woman to sit on the bare earth, floor, rafia mat or a mattress. The second "D" is what I framed as Defacement. The background to this is the belief that a woman should beautify herself to please her husband. The way she dresses, adorns herself with jewellery, wears her hair or applies her makeup is all for the pleasure of her husband, and not for herself. Therefore, when her husband dies, it is forbidden for her to do any of these things as a sign of mourning. There are appointed women called "Umuada," daughters of the lineage, whose job is to enforce a range of defacing practices. Her hair is cut, she is forced to wear black or white for varying periods of time. In some cases, widows were not allowed to bathe for up to one year; this was all part of the defacement. The third "D" is Dis-inheritance. Women were considered chattel and part of her husband's property, in the same category as furniture. Women would do back-breaking work to cultivate the family land, and yet once the husband died, she would be dispossessed of the land or house. In some instances, apart from the land grabbing, widows would be forcefully inherited by the next surviving male. This is called "levirate." This practice is also a driver for increased sexually transmitted diseases, including HIV and AIDS.

In 1995, I presented my independent research and the 3Ds concept at the Medical Women's International Congress in the Hague and was encouraged to start an NGO. When I came back to Nigeria, I was introduced by the then Director of the British Council in Enugu, Mr John Udeh to three other very dedicated women interested in widows' rights. They were Pat Okoye, a writer and publisher; Angela Otaluka a lawyer; and Patience Ifejika, a teacher and poet. Together, we started an NGO called WiDO, meaning Widows Development Organisation. The acronym WiDO was specially chosen to dispel the myth at that time that the word "widow" was bad luck and to deal with the stigma that widows suffered as a consequence.

In 1997, members of WiDO were appointed to the Enugu State Widows Welfare Committee, with my humble self as the Chairperson, by Bar Mrs Sule Ahman. We were given the task of determining the prevalent widowhood practices in Enugu State. We travelled to different parts of the State documenting the practices, those who perpetrated the practices and the community perceptions. The report of the committee showed that harmful traditional widowhood rites were still being practised in Enugu State, especially among the uneducated and rural populations.

In 1998, at the Medical Women's International Congress, I put forward an urgent statement for the elimination of obnoxious practices against widows in Sao Paulo, Brazil and this was adopted as a resolution.

In 1999, I presented a paper on the findings in Virginia, USA, at the Association for Women in Development (AWID) Forum (an international forum for women's rights).

In June 2000, I was invited to be a panellist at the United Nations General Assembly Special Session (UNGASS) in the UNIFEM Panel in New York, USA. I presented a paper on the myths, rituals and prevalent widowhood rites. In my paper at the UN, I said, "Culture is a way of life passed down from generation to generation, but when culture ceases to be a way of life and instead becomes a threat to human existence it is better defined as torture."

In February 2001, the first International Conference on Widows' rights was held in London, UK. At this conference, I presented an overview of practices from an African perspective.

I teamed up with some courageous women at the state level and carried out sensitisation and advocacy for six years to get the state law passed protecting widows. We held several community engagement convenings with traditional and religious leaders and women's affinity groups.

A major breakthrough at the local level was the passing of a bill on widow's rights by the Enugu State House of Assembly on 8th March 2001, International Women's Day, sponsored by the National Association of Women Journalists (NAWOJ). The bill was later signed into law on 28th June 2001 by the State Governor. In April of 2001, I was featured in *The Australian* newspaper after a presentation on my research into widowhood as it affects the health of widows and their children, sponsored by the Medical Women's International Association (MWIA) in Sydney, Australia. These efforts were specially recognised in a letter from the Governor of New South Wales to Governor Chimaroke Nnamani of Enugu State.

In October of 2001, I had the opportunity to present the work of WiDO to Princess Anne, the Princess Royal. In addition, with the support of the US State Department's Bureau of Education & Cultural Affairs, I facilitated a workshop for 44 Traditional Rulers in Enugu State on gender issues, including widows' rights.

In June 2003, I made a presentation at the UK House of Commons at the invitation of the Labour MP Tony Colman, representing Putney.

In an effort to ensure that the Bill on Widows' Rights is put on the national agenda, WiDO carried out a forum on legislative advocacy at the National Assembly on the 8th of March, 2004, International Women's Day, in collaboration with the Global Fund for Women.

On 23rd June 2006, I made a presentation at the Foreign and Commonwealth Office, UK, at the International Women's Day Conference, chaired by the then-wife of the UK Prime Minister, Cherie Blair, and organised by the Loomba Foundation.

The harmful practices opened my eyes to the different types of violence against women all around me but more importantly, that laws could change things. I am not alone. There is a realisation of the need for global systems change. I joined the women's movement, bonded by a passion for change and shared values. It is dangerous work, and I faced opposition from the cultural establishment.

That was what led me to take my skills and knowledge as a doctor into development and human rights work. I feel as though I am on a mission. As a clinician, I saved one life at a time, but now I find myself changing the lives of whole populations through targeted interventions and advocating for laws. At the national level, I worked with the Legislative Advocacy Coalition on Violence Against Women (LACVAW) for 14 years on the Violence Against Persons Prohibition Act (VAPP). Of course, it was by no means a one-woman show, but working together with other very dedicated people.

At the global level, I found the power of the collective voice for advocacy. I met Margaret Owen, who had started the work on widows' rights when she was asked how come she still kept her home after her husband had died. She has remained my mentor in the widows' rights movement.

My purpose is to make an impact in my generation and leave a lasting legacy by relegating violence against women and girls to history books and archives. To eradicate violence against women and girls everywhere (or in my nation, community), I envision a world where every woman and girl can thrive through lives free from violence.

When we are working on a problem as big as violence against women, the stakes are high. It's hard to count our successes when there is still so much to be done, so much violence. Yet, we must celebrate every milestone and every win in our leadership journey!

Having participated in bringing about change at sub-national and national levels, ending violence against women and girls has become my life's work. My quest for global systems change in addressing violence against women and girls has gained momentum through my involvement with the Every Woman Treaty coalition. I became co-founder of Every Woman Treaty, a global coalition of women's rights frontline activists, survivors and organisations in 147 countries, representing every region of the world, advocating for a treaty to end violence against women and girls. Globally, more than one in three women are raped, beaten or assaulted making it an existential threat of pandemic proportions. This violence is 100% preventable, and yet and unless and until we have a global treaty to end every form of violence against women and girls on earth, the violence will continue. A treaty in the form of an Optional Protocol to CEDAW, the Convention on the Elimination of All Forms of Violence Against Women, dedicated to ending violence against women and girls has become absolutely urgent and essential in order for women and girls to live a life free from violence. I invite readers to visit everywoman.org and sign the people's treaty.

The school of life is one that you enrol in without applying and without knowing what date or time you will be expected to graduate. Circumstances surrounding one's birth, choices made along the way and how much one is influenced by the environment will determine to a large extent the experiment called "the school of life."

I believe my personal experience as a child in the Biafra War, a civil war in Nigeria in the late 60s, was a motivating force that led me to work as an expert on women, peace and security. I had the opportunity of working for five years on the DFID-Funded Nigeria Stability and Reconciliation Programme (NSRP), a peace building programme managed by a consortium of the British Council, Social Development Direct and International Alert, first as a gender adviser and then as the Women and Girls Manager.

In this role, I worked to support the Nigerian Government to draft and launch their first National Action Plan (NAP) on Women, Peace and Security (WPS) 2013–2016 within the framework of the United Nations Security Council Resolution 1325 and related resolutions. Additionally, I supervised a federal-level civil society network on WPS and eight state-level WPS networks, culminating in eight state and five local government action plans on WPS. I also provided technical assistance for Nigeria's second NAP on Women, Peace and Security (2017–2020), earning me the name "Mama Peace." I went on to become the team lead and gender expert on the drafting team for the AU Stabilisation, Recovery and Resilience Strategy for the Lake Chad Basin countries affected by Boko Haram (Cameroon, Chad, Niger and Nigeria). Today, my work for peace continues as I am currently the Nigeria Co-Country Representative for the Working Group on Women, Youth, Peace and Security for West Africa and Sahel (WYPS-WAS).

I have since authored several publications and received numerous awards of merit. My personal profile is featured in the Friedrich Ebert Stiftung publication, "The Hands That Build Nigeria: Nigerian Women Role Models." In May of 2023, I was inducted into the Nigeria Women Hall of Fame. It was quite a joyful moment to see my name etched in marble.

LEADERSHIP IN HEALTH AND TECH

On the 27th, July 2019, the 2019–2022 MWIA executive acclaimed me as President-Elect.

My ascendancy to the position of MWIA President came early on account of the unexpected resignation of my predecessor on 24th November, 2019. It was heartwarming to be celebrated by my National Association, the Medical Women's Association of Nigeria (MWAN), at their National Executive Council meeting in December 2019. I was celebrated with pomp and pageantry, with a guard of honour mounted as I was ushered in with dancing and a standing ovation.

I was also warmly received by the wife of the Executive Governor of my home state, Abia State, Her Excellency Mrs Nkechi Ikpeazu.

The year 2020 started with celebrations in my honour on the rise to the position of President by the women of Saint Matthews Anglican Church, Abuja, led by Dr Chizoba Wonodi and Mrs Azuka Uwanaka, who hosted me for a dinner in January.

This was followed by another reception hosted by Professor Lovett Lawson, Professor of Tropical Medicine and Chairman of Zankli Medical Centre, Abuja, and his wife, an active MWAN member, Dr Olufunke Lawson, vice-chair and Consultant Paediatrician at Zankli Medical Centre, Abuja.

In commemoration of International Women's Day 2021, I was presented to His Excellency, Muhammadu Buhari, President of the Federal Republic of Nigeria, by Dame Pauline Tallen, the Honourable Minister of Women Affairs. I went in the company of my family and members of the Medical Women's Association of Nigeria. News had broken out about a strange virus in Wuhan, China. Little did we know that there would soon be a pandemic, and yet, in my thank you speech on that day at the Presidential villa, I encouraged the Nigerian Government to work with MWIA as a preferred partner in women's global health and urged the government to respond meaningfully to the COVID-19 pandemic as was the case for Ebola. I now have the mandate of playing a leading role as an African woman in health leadership in Nigeria and beyond.

As President, I was called upon to come up with a theme for the 2022–2025 triennium. I came up with the theme: One Humanity: Health Solutions Through Our Partnerships. As one humanity, there will be mutual respect and dignity notwithstanding our differences.

Partnerships for health solutions, therefore, constitute a global strategy highlighting the need for an approach that is holistic, collaborative and integrated.

We will invest in strategic and integrated partnerships while advocating for health solutions that are innovative and contextualised for a positive impact in each of our regions, especially addressing the issues of women's research and development.

Reinforced gender biases by health professionals towards women may lead to wrong or inappropriate diagnoses and delayed treatment. Peculiarities of women are sometimes not factored into design throughout the innovation ecosystem. Medical education programmes lack the integration of sex and gender differences in the training and retraining of healthcare professionals. According to the ILO, women are 70% of the health workforce and yet they are lacking in leadership positions.

The current tech landscape lacks consideration for the differences in physique, build, daily routine, childcare obligations of women and the investment pattern is fragmented. Women are affected differently, disproportionately and uniquely through the different phases of their lifespan, and it is crucial to recognise this in tech.

Here are some examples.

Until recently, crash tests were conducted with male crash dummies. Seat belts were not designed taking into consideration the breasts of women. Therefore, women are "47% more likely to be seriously injured and 17% more likely to die than a man if they're in a car crash."

There are only 12% of women in health tech, and few are African women. Statistics show that only 9.7% of funding goes to female-founded health tech startups, and women get smaller research grants. A Deloitte study revealed that women's health represented just 2% of the $41.2 billion in venture funding that went to health innovators.

A recent report titled "Addressing Gender Inequalities in National Action Plans on Antimicrobial Resistance" revealed women and girls face increased risks of drug-resistant infections due to limited access to clean water and sanitation, as well as their roles in household health care. Women are 27% more likely to receive antibiotics than men, highlighting biases in prescribing practices.

While the WHO Global Health Estimates of 2019 suggest that men are more likely to die by suicide than women, studies have found that women are more likely to self-harm and attempt suicide due to gender-based violence, economic inequality, discrimination and lack of access to mental health services.

Despite the limited gender-disaggregated data on the prevalence of certain skin non-communicable diseases (NCDs), preliminary indications suggest a higher disease burden and stigma in women, including skin-disease-related mental health issues (in particular depression, suicidal ideation and anxiety) and access to work.

Having identified the problems, multidimensional solutions must be clearly outlined.

There should be deliberate effort to mentor women to enter and stay in STEM.

Advocacy should be stepped up to influence policies to fund women-led tech innovations and ventures with a gender-intentional focus.

Data diversity and integrity are crucial so as not to create a false sense of precision while at the same time accommodating outliers.

Machine learning and AI in medicine are encouraged with the use of drones to deliver to hard-to-reach communities.

Digital inclusive and safe environments are to be established where interventions should do no harm. We also need to be aware of the risks, including tech-facilitated stalking.

Innovators must ensure safety by design and remove safety biases.

Digital health innovators that affect health outcomes must be deployed equitably and cost-effectively.

The design of algorithms should be done through meaningful and inclusive consultations.

Because AI has a voracious appetite for energy consumption, carbon-sensitive alternatives must be explored.

IN CONCLUSION

As a woman of faith, I view my ascendancy to the position of International President as a beautiful and divinely orchestrated gift from God. I am profoundly grateful, humbled and immensely elated.I have also felt a huge sense of responsibility.

Health leadership through my leadership of the Medical Women's International Association (MWIA) has provided me with opportunities for peer-to-peer and intergenerational mentoring, participation in high level advocacy spaces, as well as pursuing partnership opportunities and better understanding better the benefits and risks of tech enabled and tech leveraged health.

Despite being born at a time when there was a culture of son preference as earlier stated, on account of my achievements, rising above cultural inhibitions, my community and state have conferred on me the honorific titles of:

Ada e ji aga mba' – Daughter who represents her people in foreign lands – 2007.

Odo – Abia – The Adorning and bright ornament of Abia – August 2014.

Osi na ulo mara mma – starting at home to do good globally – September 2021.

As a woman in health leadership, I am making a clarion call to action for more investment in innovative solutions and resources in addressing disparities in health leadership and issues that disproportionately affect women. In a male dominated area, the mindset should be to overcome challenges of gender, culture and colonial legacies and to reflect women's unique realities.

My favourite saying is "encouragement is like sunlight to a closed flower." So I try to encourage those I come across daily by speaking words to uplift them with grace.

Together we can amplify our voices, and elbows locked, we can form a formidable barrier in order to face any challenges that may come our way.

Poet Maya Angelou said, "The universe is not made of atoms but stories."

This is my story, her story!

6 Agile, Emerging Technologies, and the Future of Women-Led Innovations in STEM

Stella Ihenacho

INTRODUCTION

In 2018, Amina Bakari, a software engineer from Kenya, was leading a tech project to develop an e-learning platform for rural communities. Having to balance a team spread across different time zones, along with tight deadlines and severely limited resources, traditional management methods just weren't cutting it. Delays and miscommunication with the team were holding up progress. Determined to find a solution, Amina turned to Agile methodologies and focused on Scrum frameworks that could bring her team together in a more collaborative and adaptive way. By adopting Agile, Amina not only delivered the project on time but also set a new benchmark for tech projects across her organization—demonstrating how a flexible, iterative approach could reformat not only the outcomes but also the working culture itself. Her success with Agile didn't just complete one project; it opened doors for her and other women in the organization, showing that Agile could indeed empower women to overcome systemic barriers and lead complex technological initiatives with confidence.

Amina's life is but one of many tales that reflect the reality of African women who use Agile to be contributors and leaders in the STEM fields. The intersection of Agile with emerging technologies has created a most empowering space for African women in STEM; a space that furnishes them with the tools and frameworks necessary to drive innovation and impact economic transformation across the entire continent. But this path is often met with obstacles, from learning to work in male-dominated workspaces to overcoming societal stereotypes about women in tech. In these areas, Agile methodologies are removing the barriers by offering collaborative and adaptive environments where African women can thrive, lead, and inspire the next generation.

This chapter discusses how Agile practices, with their flexibility and emphasis on teamwork, are well-suited to support African women in STEM. Agile is more than a means for managing projects; it is a mindset, a methodology that values quick

 DOI: 10.1201/9781003540618-6

adaptation, open communication, and continuous learning—features vital to succeed at the ever-innovating, fast-moving landscapes of most technological fields. For women, who are usually underrepresented in STEM, Agile provides a structure that allows them to work around the traditional hierarchies and silos, thus opening opportunities for influence and growth in careers that sometimes could have been inaccessible before. Agile helps African women bridge the gap between innovative potential and real-world impact, addressing structural inequities so often pervasive in the STEM fields.

This will also look at how emerging technologies, such as artificial intelligence, blockchain, and the Internet of things, are reshaping Africa's economic and social landscape. These technologies, when combined with Agile, are opening up pathways for women to lead projects, solve complex problems, and deliver sustainable solutions to some of the most pressing challenges across the continent of Africa. For example, women in fintech and data science can participate in the development of blockchain technology to create transparent, secure financial systems that will be able to support economic growth and increase financial inclusion for underserved communities. Similarly, women-led artificial intelligence projects are making tremendous progress in healthcare, where AI-driven diagnostics and data analysis make quality medical care possible even in the most remote regions.

In this chapter, we will look at case studies of African women who have adopted Agile to lead transformative STEM initiatives—demonstrating the power of this approach in real-world applications. By reading these case studies, one will see the tangible changes Agile methodologies and emerging technologies have made in their careers and larger communities. The chapter will also look into specific ways in which Agile has allowed these trailblazers to better navigate cultural and societal hurdles, enhancing their ability to succeed in roles traditionally dominated by men. With these examples, this chapter tries to underscore the important role Agile can play in leveling the playing field for women in STEM and providing them with the tools they need to lead innovation and economic growth.

This chapter ultimately looks to an African future in which women in STEM are not just empowered to participate in, but shape and lead the continent's technological evolution. Agile methodologies provide a framework in which women can engage in meaningful and impactful work to advance their own growth and Africa's development into becoming a global tech leader. Combining the principles of adaptability, transparency, and collaboration that Agile brings with it, African women are rewriting what it is to be an innovator and leader in the tech industry by tapping into emerging technologies.

The confluence of Agile and emerging technologies brings about unprecedented opportunities for African women to break stereotypes, innovate, and drive lasting change. This chapter is meant to be a blueprint and an inspiration for those interested in understanding and supporting the pivotal role African women in STEM play in propelling Africa's economic advancement. With Agile, they're building not just careers but the foundations for a future wherein women-led innovation is at the core of Africa's technological and economic transformation.

BACKGROUND INFORMATION

Evolution of Agile in Technology Sectors

The Agile methodology was created in the early 2000s as a response to the rigid, often slow-moving traditional project management practices. The Agile approach transformed how teams in technology and software development approached their work. Based on the principles outlined in the Agile Manifesto of 2001, Agile came about as a framework meant to enhance adaptability, collaboration, and efficiency in an increasingly complex technological world. Unlike the traditional, linear approaches to project management—and the Waterfall model in particular—Agile focuses on iterative development. In this regard, teams split up the work into smaller, manageable cycles called "sprints." Each sprint then delivers a functional segment of the project, enabling the team to review progress made, take on board feedback, and make any changes that are needed.

Agile has become such a visible reality due to its flexibility; it allows development teams to change direction quickly in response to changing requirements, customer feedback, and shifting market conditions. Starting in the realm of software development, Agile has begun to find widespread application in other segments, such as healthcare, finance, education, and even public policy. The iterative nature of Agile makes it well-suited for environments that face rapid technological changes, which is often the case in today's technology-driven economy. By focusing on flexibility, Agile encourages an innovation-friendly environment where team members are empowered to try and fail, to learn from mistakes, and to find better ways of doing things.

Agile has also developed to integrate a variety of frameworks, such as Scrum, Kanban, and Lean, each with its own set of practices and tools that help team collaboration and efficiency. For example, Scrum focuses on defined roles such as Scrum Master and Product Owner to coordinate activities within the team, ensuring that everyone has the same vision of the project goals. On the other hand, Kanban deals with the visualization of workflows and constraints on work in progress so that the team is able to manage resources optimally. These frameworks empower teams to approach problem-solving creatively and adaptively, driving innovative solutions that meet the demands of an increasingly complex, interconnected world.

Agile's reach has also extended beyond the project level and into organizational transformations. Nowadays, numerous organizations apply "Agile transformations" in the course of trying to ingrain Agile principles into the organizational culture, structure, and strategies. This represents a change that is quintessential in establishing an environment of innovation, where teams are self-organizing and leaders support continuous learning. In Africa, Agile has been an important methodology that helps organizations compete globally. It gives them a structure to increase productivity, empowering diverse teams, and responding to local and international market needs. What is most important for African women in STEM is the cultural change Agile fosters: it is one that levels the playing field, breaks down traditional hierarchies, and allowing skills and innovation to rise above rank and tenure.

Role of Women in STEM in Africa

For centuries, African women have been trying to participate in the development of the STEM fields on the continent. One has been able to note the societal, educational, and professional barriers in this respect. In the past, women in Africa have encountered other major entrance barriers into the STEM fields, including inadequate access to quality education, gender biases, and sociocultural expectations that shape career choices. For instance, it has been reported that girls in Africa often shy away from science and mathematics due to cultural stereotypes that these disciplines are too difficult or inappropriate for women. Therefore, there exists a large gender gap in most STEM fields across Africa, where women are underrepresented in higher education and professional settings.

Though these are some of the challenges, African women have been making tremendous strides in STEM, most often initiating initiatives and innovations that address both local and global issues. Organizations, NGOs, and even governmental policies have stepped forward to support women in STEM in an attempt to tear down structural barriers and create opportunities for women to pursue and succeed in STEM careers. Such initiatives as the African Union's Science, Technology, and Innovation Strategy for Africa—STISA-2024, along with local programs like Nigeria-based "She Codes Africa," are working to promote STEM education and opportunities for women, encouraging them to take up and remain in STEM careers.

Yet, significant challenges remain. African women in STEM often face workplace biases that manifest in unequal pay, limited career advancement opportunities, and minimal representation in leadership roles. The male-dominated nature of many STEM fields can create environments where women struggle to find mentorship, advocate for themselves, or challenge ingrained stereotypes. This therefore brings about a lack of role models for the younger women seeking to get into these fields. For Africa's technological and economic growth to really gain from the capabilities of its entire population, it needs to target these gender-based disparities and make sure that women in STEM are working in inclusive, supportive environments where they can thrive.

The Intersection of Agile, Emerging Technologies and Gender Equality

The confluence of Agile, new technologies, and gender equality offers an unparalleled opportunity for women in STEM, especially in Africa. Agile practices, with their traits of collaboration, flexibility, and empowerment, are therefore very befitting in supporting women who often face gender-specific barriers in their careers. Agile methodologies, unlike traditional, hierarchical structures, which reinforce power dynamics, focus on a team-based approach in which every member's input is valued and thus create a more inclusive environment for women. In this way, by promoting the role of Scrum Master and Product Owner, Agile offers women a chance to attain leadership positions in project management and lead teams toward project goals without the need for traditional hierarchies of management.

Moreover, Agile methodologies align quite well with the speed of innovation in emerging technologies. Artificial intelligence, machine learning, data science, and blockchain are all very fast-paced fields; there is a need for adaptability and a quick response to changes. The iterative approach of Agile gives women in these sectors a framework to experiment, learn, and evolve alongside technological developments. This synergy between Agile and emerging technologies empowers African women in STEM to contribute not only to innovation but also to lead it, breaking down traditional barriers and forging new paths across tech-driven industries.

By addressing and neutralizing some of the biases inherent in traditional organizational structures, Agile provides a valuable framework to promote gender equality in tech. In Agile, for instance, a focus on measurable outputs and regular feedback loops can mitigate biases that more often than not influence performance evaluations. In Agile, performance is judged by contributions toward team goals and not by subjective criteria—a good way to create fairer assessment systems. Furthermore, the collaborative nature of Agile encourages mentorship and sharing knowledge, which are great career-growth and support-network components, especially in male-dominated industries. This approach creates an environment where women are more likely to be recognized for their skills and contributions, as well as find opportunities to develop their careers.

The intersection of Agile, emerging technologies, and gender equality also supports African women in STEM in navigating societal expectations and building up resilience against biases. In an Agile environment, it can be much easier for women to balance their career aspirations with other responsibilities, as the frameworks in Agile value outcomes over physical presence or strict observation of time. The increase in the diversity of the workplace makes this a possibility. This flexibility might prove highly important for women in STEM who carry extra responsibilities away from work. Agile methodologies, therefore, create a support structure for women, considering diversified backgrounds, experiences, and needs, in order to enhance their potential for being successful in STEM careers.

African women in STEM are sure to be leading the way in this transformation, as emerging technologies continuously shape Africa's economy. By using Agile, they will not only ensure project success but also advance themselves in a tangible way. As African women adapt to new demands arising from emerging technologies, they drive innovations that have lasting social and economic impacts across the continent. Adaptability is essential in fields where technologies are rapidly changing and require continuous learning and flexibility. African women can use Agile to overcome societal and professional barriers to leadership, create gender-inclusive environments, and contribute to growing the African economy by using it.

MAIN CONTENT

Agile Methodologies for Inclusive Innovation

Agile methodologies provide a strong platform from which inclusiveness and innovation, with special emphasis on women in the environment of STEM, can be propagated. Unlike the more traditional types of project management, Agile provides a very collaborative and adaptive framework where each member within the team is welcome to contribute, hence giving rise to various views and skill sets. This is important for women, who generally bear the brunt of systemic barriers to entering male-dominated industries, as Agile enables them to take on leadership roles and active participation in decision-making. In this regard, Agile's emphasis on iterative progress with transparency and accountability creates an environment where women in STEM have the confidence to put forward ideas, try out new concepts, and make adaptations accordingly, free from punitive judgment.

Agile's framework is especially helpful for women-led STEM initiatives because its principles have to do mainly with the aspect of self-organization, flexible work structures that align with the diverse range of women's personal and professional responsibilities. For example, a Scrum environment allocates defined roles, such as Product Owner or Scrum Master, where women can be given responsibility for the project vision to guide and steer teams toward successful projects. This, in turn, will give them the opportunity to demonstrate leadership skills in technology projects. Moreover, Agile stresses quantifiable results rather than hierarchical structures, allowing women to prove themselves based on performance and thereby creating a clear career path toward professional recognition.

Agile also feeds problem-solving and adaptability skills, which are in particular demand in STEM, as technology and requirements constantly change. Scrum, Kanban, and related Agile frameworks allow for iterative feedback and adjustments, thus helping women in STEM make quick turns and solve challenges in near-real time. By building a fail-forward culture, Agile reduces the stigma of experimentation, thus making it easier for more women to innovate, take risks, and create new solutions. This flexibility of Agile, especially in areas such as software development, data science, and engineering, allows women to lead teams that answer the pressing demands of a rapidly changing, technology-driven economy, thereby helping to contribute to the causes of economic growth and technological advancement.

Case Studies of Women-Led Initiatives

With the application of Agile methodologies, women all over Africa are driving innovative, transformational initiatives in technology, education, and non-government organizations. The case studies that follow serve to illustrate just how Agile empowers and equips African women to drive projects, overcome obstacles, and craft economic and social impact.

Case Study 1: Woman-Led Technology Startup Applies Agile

One such example is Tech Sisters, a Nairobi-based technology startup designed by, and for, women to address local problems using digital solutions. Applying Agile principles, Tech Sisters develops software applications that can be used to extend information and resource access in the education and health sectors within disadvantaged communities. Agile frameworks, in particular Scrum, ensure that the team quickly designs, deploys, and re-issues the digital tools at extremely short intervals.

The team integrates the changes in productive biweekly sprints to keep their software responsive to the needs of the community.

The leadership of Tech Sisters has successfully adopted Agile to create a collaborative, inclusive work culture that attracts female engineers, designers, and data scientists. Guided by the foundational Agile principles of teamwork and adaptability, the founders have developed an enabling environment where free-flowing dialogue is encouraged and continuous learning is fostered. This inclusive culture has reduced the gender gap within their workforce and encouraged more females to take up STEM careers in Kenya. This was an avenue that saw Tech Sisters gain international NGO funding to expand projects and reach out to more people due to these efforts—a sure testament that Agile drives innovation, not just professional growth for women-led initiatives.

Case Study 2: Educational Programs for the Promotion of Agile-Driven STEM Careers Among African Women in the Diaspora

In addition, educational programs outside of Africa also use Agile methodologies to promote STEM careers among African women in the Diaspora. Operating in Atlanta, Georgia, is NSTAR Technologies, a training institution that trains African women in the diaspora with the aim of preparing them for the future of work and access to opportunities in STEM careers. NSTAR Technologies uses Agile frameworks in structuring its programs; therefore, the curriculum changes with time, as influenced by the changing tech landscape and reviews by participants. This iterative training program is composed of modules such as process improvement and Agile methodology, where students learn how to apply Agile in simulated project environments and proceed to get opportunities in Fortune 500 companies.

NSTAR Technologies further equips young women with project management and teamwork, which are skills important in the technology-driven economy of today. The Agile approach helps them deal resiliently with real-world issues, thus transitioning them into taking on projects in technology. Finally, NSTAR Technologies graduates emerge much more ready for software development, data science, and project management; many apply these to their businesses, and some create their own venture. In this light, NSTAR Technologies has successfully fashioned itself as a pipeline of female STEM talent to help meet the greater mission of reducing gender disparities in STEM fields.

New Emerging Technologies Shaping Women-Led STEM Initiatives

New technologies, including AI, IoT, and blockchain are now redefining STEM with exciting careers for African girls and women who drive change as leaders and innovators using the Agile way of working. AI presents fundamental applications, for example in health care, agribusiness, and monetary innovation; key drivers of financial growth all over Africa. Agile methodologies applied will enable women in STEM to build incrementally the AI-based solutions for predictive healthcare models and smart farming tools, which can be continuously improved and validated throughout the process with stakeholders.

The other space is in the evolving financial technology landscape where blockchain technology also provides an opportunity for women leaders in Africa. By integrating Agile, women-led fintech startups are empowered to build decentralized types of financial solutions that promote transparency, accessibility, and security.

Agile is an iterative method which allows for the startup to test its blockchain application, get feedback, and provide services in layers. This would ensure that every release will actually be up to user needs and regulatory standards. This gives a chance to women entrepreneurs to create access to financial services in the unserved regions in Africa, and this most critical function of promoting economic empowerment and financial inclusion

In IoT, it empowers Agile even more, allowing women of STEM to design and write software that actually reacts to real-time data where it is being produced—like remote monitoring in agriculture or energy management. It would also fix problems unique to African American lives using adaptive and data-driven solutions. For instance, IoT-powered smart sensors continuously monitor the soil quality to update African female agronomists with the same information. This, in turn, will help them arrive at a decision to support sustainable farming practices. On the other side, Agile methodologies are useful in enabling professionals to configure IoT applications in a way that performs on par with the changes in the environment for the efficiency and sustainability of their solutions in the long run. These rising technologies juxtaposed with Agile place African women at the forefront of digital transformation, thus enabling them to reduce all gaps in technologies and take lead positions in the most important sectors.

Agile Mindsets as Career Growth Accelerators

This is the Agile mindset: continuous improvement, collaboration, and adaptability or a culture which significantly fuels the growth of careers for women in STEM much faster. Agile emphasizes learning by doing, thus allowing women to gain credibility, build their skills, and attain leading roles with much more ease compared to traditional career paths, which may have hindered them to do so. For instance, the continuous improvement principle enables women to regularly assess their situation, set incremental goals, and adopt lifelong learning that helps them to stay relevant with respect to emergent areas in technology.

Collaboration is another important principle of Agile, which again can play the role of a means for women in career development. Agile teams encourage their members to share knowledge, facilitate others' progress, and work together toward one common objective. This can be very complementary to the competitive, individualistic approaches and can forge a great support network for women in STEM. Working in Agile teams, the visibility for women increases, mentorship ensues, and it is the relationships that are formative in professional development. Also, the collaborative nature of Agile provides an avenue to have women in leadership in projects and to lead team discussions through their communication and decision-making capabilities, which are critically important attributes that will equip them for leadership.

The ability to adapt easily is one of the main tenets of Agile. Therefore, Agile helps women to gain confidence through the changing demands of technology. Agile can keep the professional focused on any change of direction or course; a mindset particularly helpful for women in the fast-moving STEM world. Thus, the ability to adapt to new technologies, project scope, and market trends enables women to capitalize on emerging opportunities and high-impact roles. This can also allow women to overcome some of the obstacles such as poor access to career growth or gender biases in environments that foster innovation and flexibility rather than traditional hierarchical constraints.

The Agile mindset, therefore, affirms women in STEM with career advancement opportunities, not only by piercing systemic barriers but eventually building themselves up into leaders in Africa's growing technology sector.

DATA ANALYSIS: ANALYSIS OF CASE STUDY DATA

Analyzing data from the case studies presented offers valuable insights into how Agile methodologies influence women-led advancements in STEM across Africa. The case studies of *Tech Sisters* and NSTAR Technologies *(NSTAR Tech)* demonstrate measurable improvements in productivity, innovation, and inclusion due to the adoption of Agile practices. By examining specific metrics, including project completion rates, team diversity, and innovation output, we can gauge the broader impact of Agile on career advancement and gender equity in technology sectors.

Project Completion and Iteration Success Rates

One primary indicator of Agile's effectiveness is the increase in project completion rates and iteration success among women-led initiatives. In NSTAR Tech's case, the use of Scrum sprints resulted in a project completion rate of 90% across their core applications, with an average sprint duration of two weeks. This rapid iteration cycle allowed the team to respond to user feedback efficiently, achieving higher-quality outcomes in less time. Agile's emphasis on short, iterative development cycles also helps women-led teams avoid common pitfalls in traditional, linear project management, such as extended timelines and scope creep. Agile's adaptability ensures projects remain aligned with user needs, fostering a sense of accomplishment and growth within teams, a crucial factor in career development.

In NSTAR Tech, Agile's structure is beneficial in delivering technical education. The program's iterative module design ensured a retention rate of 85% among participants, significantly higher than the average retention rates of traditional STEM programs. The Agile framework allowed NSTAR Tech to revise and adapt the curriculum based on participant feedback, making it more relevant and accessible. For women pursuing STEM careers, this model demonstrates how Agile can improve educational outcomes, paving the way for increased career entry and advancement opportunities.

Team Diversity and Role Distribution

Data from the above also indicates that Agile encourages inclusive team structures, particularly in terms of role diversity. Within NSTAR Tech., 70% of students hold leadership or decision-making roles, including positions as Product Owners, Scrum Masters, and Agile Coaches. This distribution of responsibility contrasts with more hierarchical structures, where decision-making power is typically concentrated at the top. Agile's decentralized approach empowers women to lead specific project components, promoting a culture of equality and skill development. Agile fosters diverse contributions by emphasizing collaboration and shared accountability, which supports an inclusive work environment where women feel motivated to take on leadership roles.

Additionally, NSTAR Tech's Agile-oriented training led to diverse role distribution within its project teams. Participants gained hands-on experience in Agile roles, learning to manage tasks as Scrum Masters and Product Owners. This diversity in role experience significantly increases young women's employability in STEM, where role versatility is often valued. Through this structure, NSTAR Tech has successfully equipped its participants with the skills necessary for Agile environments, allowing them to transition into tech roles more smoothly. Data indicates that NSTAR Tech graduates are twice as likely to secure entry-level roles in tech companies as their non-Agile-trained peers, highlighting Agile's role in building career-ready, versatile professionals.

Innovation Output and Scalability

Agile's focus on iterative feedback and continuous improvement directly impacts innovation output in women-led STEM initiatives. Tech Sisters, for instance, developed three major applications in health and education within 18 months, each iteration incorporating community-driven insights and enhancements. The Scrum process enabled rapid prototyping, leading to applications that received high user satisfaction ratings and continued funding support. Agile's iterative model also encourages scalability; as user needs evolve, projects can easily expand to meet new demands, allowing women-led startups like Tech Sisters to scale their impact across broader regions without compromising quality.

In NSTAR Tech's case, Agile's adaptive framework enabled the organization to develop tailored programs that meet the unique needs of different African women. Agile's scalability has allowed NSTAR Tech to replicate its programs, reaching over hundreds of girls and women globally. This expansion demonstrates Agile's potential to create scalable, women-led educational initiatives that adapt to regional nuances, thereby enhancing the accessibility of STEM education for girls across diverse communities.

Empowerment and Skill Development Metrics

Finally, qualitative data from these case studies highlight the empowerment and skill development outcomes Agile provides for women. Feedback from NSTAR Tech. participants reports a 70% increase in confidence regarding their technical and project management skills, attributable to Agile's hands-on, iterative learning approach.

In summary, data from these case studies underscore Agile's transformative impact on women-led STEM initiatives, promoting higher completion rates, diverse leadership, increased innovation, and measurable skill development. Agile not only enhances productivity but also fosters an inclusive, empowering environment that enables African women in STEM to thrive and drive meaningful change.

FINDINGS/CONCLUSION

The exploration of Agile methodologies within the context of African women in STEM underlines Agile's transformative power, which catalyzes empowerment, career advancement, and innovation. The chapter emphasizes how Agile frameworks provide an enabling environment that thrives on adaptability, inclusivity, and rapid problem-solving—most relevant to facing Africa's dynamic and usually unfriendly technological landscape. In many instances, Agile affords a reasonable chance for African women to rise into positions of authority and influence in an industry dependent on STEM that is usually very male-dominated, thus opening more avenues for personal and professional growth.

The iterative approach to continuous improvement by Agile goes hand in hand with the high speed of technologies such as artificial intelligence, blockchain, and IoT. Agile practices equip women in STEM with the agility needed to adapt quickly to such technologies for tuning up solutions through relentless feedback and collaboration. This will make them great, confident professionals, innovators, and leaders across industries. Agile contributes to an even deeper digital transformation that cuts across Africa, more inclusively so, because it allows a more active participation of women in the creation and deployment of emerging tech solutions.

Agile methodologies also provide a work environment where African women in STEM can join and work together by leading others while sharing and, at the same time, encouraging the growth of their peers. Emphasis on collaboration would therefore break down the traditional hierarchies of the workplace and provide women with visible roles in decision-making and project management. Agile's structure thus provides a venue for women to show their competency and apply for leadership positions that have been traditionally held by men. In this way, they can contribute to solving societal and technological problems. One can see it in practice in various cases, such as Tech Sisters and NSTAR Technologies, where women use Agile for social change, networking, and also to inspire other girls to be part of STEM.

To conclude, Agile holds great power in guaranteeing gender inclusivity and stirring sustainable innovation in the STEM sectors of Africa. Agile fosters adaptive, inclusive work cultures that help empower African women in the area of STEM, thereby enriching the African economy with multiple perspectives and solutions. With the continued adoption of Agile methodologies by women-led initiatives and tech startups, Africa is set to take the next leap as a global leader in innovation through the resilience, creativity, and leadership of its women in STEM. The findings of this chapter represent Agile as not only a methodology but a driver for social and economic progress, enabling women across Africa to break down the barriers, transform industries, and inspire a new generation inclusive of technological progress.

DISCUSSION

With the rise of Agile methodologies throughout Africa, there is an increasing impact on policy frameworks and STEM education. This could be instrumental in leading government and organizational policies toward more inclusivity regarding gender, equal access to technological education, and professional growth. It's in the incorporation of Agile principles into STEM policies that educational institutions create more supportive, Agile learning environments, preparing women for leadership in innovation.

Most African countries do not currently have STEM-specific policies that address the gender gap or support women's entry into high-demand, Agile-driven roles. Agenda items on education and training programs can integrate Agile to help close these gaps; they provide skills training beyond just technical know-how. Agile, with its strong tenets of adaptation, collaboration, and continuous improvement, places women not only at an advantage for entering but also for advancing in most STEM fields. For instance, Agile training at technical schools and universities would immediately raise women's readiness to enter those careers in emerging technology sectors. Besides, educational programs on the Agile mindset can instill in young women an entrepreneurial spirit to set up and run tech startups that will bring innovative solutions to their own communities.

On the educational front, this may be realized by creating linkages between government agencies, private technology firms, and academic institutions. Such collaborations would allow for Agile workshops, internships, and mentorship activities specifically targeting young women. It would also help education-focused NGOs and community-based organizations train women in rural and underserved communities in using Agile methodologies, hence offering career opportunities that may otherwise be out of reach. Such initiatives could be a landmark in bridging the urban-rural divide and increasing access to quality education and jobs in STEM for women.

FUTURE DIRECTIONS

Looking ahead, Agile combined with emerging technologies like AI, blockchain, and IoT holds immense promise for advancing African STEM fields. As these technologies keep changing the face of industries like healthcare, agriculture, and finance, Agile principles can be applied to encourage women's participation in the development and management of cutting-edge solutions tailor-made for Africa's unique challenges. For example, Agile's iterative approach could help women-led teams in refining AI-driven applications that optimize crop yields and enhance food security across the continent. Similarly, the use of IoT by Agile teams in health could create decentralized health monitoring systems where women are leading the projects to provide affordable and accessible health solutions in remote areas.

Agile also opens up scalability for women in STEM to take on leadership positions in projects on environmental sustainability and renewable energy. With rising interest in this arena, Agile-driven projects may just lead to technological progress contributing to Africa's green economy. By empowering women to lead in sectors where gender representation is limited, Agile can drive a paradigm shift in creating an inclusive STEM landscape globally.

In the final analysis, Agile's future in Africa's STEM sectors looks bright, and so do the opportunities for women to shape both technology and policy. This will just need more support from governments, private organizations, and educational institutions for easy access to Agile training and equal representation. With the adoption of Agile, first in policy and then in educational frameworks, Africa will develop a STEM ecosystem where women are not just participants but leaders in creating innovative solutions to inspire future generations. As Agile continues to evolve with emerging technologies, African women in STEM are in a position to harness their power by charting new paths forward.

SOCIAL/PRACTICAL IMPLICATIONS

Incorporating Agile principles into STEM education and the workplace can transform the lives of young women and professionals in Africa. Agile methodologies have become essential for technological innovation and problem-solving, and policymakers, educators, and organizations have an opportunity to create inclusive, empowering environments for women. Embedding Agile into STEM curricula and fostering Agile-friendly work environments will support women's success and leadership in technology.

Educators and policymakers

The Agile principles provide a flexible, collaborative approach to learning that can be particularly beneficial for young women pursuing STEM careers. From early childhood education to higher education, policymakers and educators should implement Agile frameworks to equip students with the skills they need to succeed in today's fast-paced job market. Agile's iterative learning approach, which emphasizes feedback and continuous improvement, can transform educational models into adaptable frameworks where students engage deeply with technology and learn from real-world applications.

It is a practical step for policymakers to partner with tech firms and Agile-certified organizations to provide high-quality Agile training, internships, and mentorship programs for young women. Students would gain hands-on experience, develop teamwork skills, and become familiar with project-based learning that mimics real-life scenarios through these programs. Schools could also encourage participation in Agile-based extracurricular clubs where students work on tech projects collaboratively, with a focus on problem-solving, communication, and adaptability.

Educators can incorporate Agile principles into STEM by restructuring coursework to include more project-oriented tasks, team-based learning, and adaptive assessment techniques. Instead of focusing solely on final exams, instructors could introduce mini-projects and milestone evaluations to allow students to receive constructive feedback and iteratively improve their work. This approach aligns with Agile's emphasis on continuous learning and would not only enhance students' technical skills but also build confidence and resilience—qualities essential for women navigating STEM careers.

For STEM Organizations

For STEM organizations, creating Agile-driven workplaces can enhance women's experiences and opportunities for growth in tech. By establishing an Agile culture that prioritizes flexibility, collaboration, and inclusivity, companies can foster environments where women feel empowered to lead, innovate, and contribute meaningfully. To achieve this, organizations should provide Agile training for employees and encourage team structures that embrace diverse perspectives and skill sets.

A strategy for fostering an Agile-cultured environment is to implement mentorship programs where women in tech can learn from experienced Agile practitioners. Mentorship encourages knowledge sharing, offers valuable guidance, and helps women navigate workplace challenges in male-dominated sectors. Companies can also create "Agile Champions" roles within teams to lead Agile practices, such as daily stand-ups, sprint planning, and retrospectives, where all team members, regardless of seniority, have a voice. This can promote a culture of empowerment and inclusivity.

Moreover, organizations should encourage open feedback loops, a principle of Agile, that provides team members with regular opportunities to share ideas and solutions. Open communication channels and feedback systems not only enhance project and product outcomes but also give women a platform to demonstrate their capabilities and take on visible, impactful roles.

Finally, agility is also intertwined with flexibility, and flexible work arrangements align with supporting women's diverse needs. Allowing remote work, flexible hours, and part-time options helps retain female talent in STEM fields and shows an understanding of the different work-life balance needs women may have.

CONCLUSION

Integrating Agile principles into educational materials and workplace practices is essential for creating pathways to success for African women in STEM. By promoting Agile's adaptive, collaborative, and feedback-driven framework, policymakers, educators, and organizations can actively contribute to a more inclusive tech landscape. Agile-focused curricula empower young women to approach learning with curiosity and resilience, while Agile-friendly workplaces foster supportive environments where women in tech can lead and thrive. As these practices become more widespread, Africa's tech industry can expect an interest and a rise in innovation and progress, driven by women who are empowered to redefine what it means to lead in technology.

REFERENCES

Books

Beckett, S. T. (2000). *Science of -chocolate.* Royal Society of Chemistry.

Rigby, D. K., Sutherland, J., & Takeuchi, H. (2016). *Doing agile right: Transformation without Chaos.* Harvard Business Review Press.

Crawford, M., & Unger, R. (2019). *Women and gender: A feminist psychology.* McGraw-Hill Education.

Journal Articles

Sohail, F., Qureshi R., Naseem M., & Zia S. (2021). The impact of Agile methodologies on software development team dynamics. *Journal of Software Engineering, 4*, 154–157.

McDermott P., & Mack K. (2014). Gender equity in STEM, 4-29.

UNESCO. (2020). *Closing gender gaps in Africa's STEM sectors: A policy perspective.* Retrieved from https://www.unesco.org/africa/stem

Online Sources

United Nations Educational, Scientific and Cultural Organization (UNESCO). (2022, March 15). *Promoting STEM education for women and girls in Africa.* Retrieved from https://www.unesco.org

International Labour Organization. (2023). *Gender equality and women's empowerment in the world of work: STEM focus.* Retrieved from https://www.ilo.org

Reports

World Economic Forum. (2021). *Global Gender Gap Report 2021.* Retrieved from https://www.weforum.org/reports/global-gender-gap-report-2021

McKinsey & Company. (2020). *Women in the Workplace 2020.* Retrieved from https://www.mckinsey.com/reports/women-in-the-workplace

Publications on Gender and Economic Innovation in Africa:

African Development Bank. (2021). *Empowering African women in science and technology: A pathway to economic resilience.* Retrieved from https://www.afdb.org

World Bank. (2022). *Accelerating Women-Led Innovation in STEM in Africa.* Retrieved from https://www.worldbank.org

Sources on Agile and Technology for Education Empowerment:

Michael Noonan, Yi-Wang, P., Wong, W., & Fernandes, V. (2022). Leading through crisis: An agile mentality for education organisations. 8–22.

7 African Women Trailblazers in STEM

Impact of Emerging Technology—A Consumer's Perspective and the Role of NGOs

Chiso Ndukwe-Okafor

INTRODUCTION

African women are breaking barriers and redefining their roles across various sectors, with the technology space being a notable transformation arena. In the vibrant streets of Lagos, Nigeria, a young innovator is working on a smart renewable energy solution to tackle the nation's longstanding power supply challenges. Thousands of kilometers away, in a rural Kenyan village, a female farmer uses a mobile app to access real-time weather updates and expert crop management advice. Meanwhile, in a state-of-the-art lab in Kigali, Rwanda, a brilliant female scientist leads a team developing advanced artificial intelligence algorithms to address Africa's unique challenges.

These scenarios, once the stuff of science fiction, are now everyday realities across the African continent. Artificial Intelligence (AI), mobile technology, renewable energy, and other innovations are reshaping the landscape of African nations, offering unprecedented opportunities for growth, development, and an improved quality of life.

However, as we marvel at these technological advancements, we must also confront a stark reality: the persistent lack of women in Science, Technology, Engineering, and Mathematics (STEM) fields, particularly in Africa. In spite of making up half of the continent's population, African women remain a minority in STEM careers, with their invaluable perspectives and talents often overlooked or underutilized (AUDA-NEPAD, 2022).

This chapter explores the intersection of emerging technologies and the role of African women in STEM, viewed through the lens of the everyday consumer. We will delve into how these technologies are changing the lives of millions of Africans,

DOI: 10.1201/9781003540618-7

the challenges and opportunities they present, and the crucial contributions of African women trailblazers who drive innovation and progress.

By adopting a "consumer's perspective," we seek to understand how these technological advancements impact daily lives across the continent. From mobile banking revolutionizing financial inclusion to e-health solutions bringing medical care to remote areas, the effects of these innovations are far-reaching and profound. Yet, as we celebrate these advancements, we must also critically examine their implications, ensuring that the benefits are equally distributed and potential drawbacks are addressed.

As we embark on this exploration, we invite you, the reader, to consider your experiences with technology in Africa. How have emerging innovations changed your life? What challenges have you faced in adopting new technologies? And, crucially, how can we work together to create a more inclusive tech landscape that harnesses the full potential of Africa's women in STEM?

Join us as we uncover the stories of remarkable African women breaking barriers in STEM fields, examine the transformative power of emerging technologies, and envision a future where innovation and inclusivity go hand in hand. Through this journey, we aim to inspire a new generation of African women to pursue STEM careers, encourage consumers to support women-led tech initiatives, and contribute to building a more equitable and technologically advanced Africa.

AFRICAN WOMEN IN STEM: BREAKING BARRIERS

In the heart of Cameroon's tech scene, Theo Nyenti stands as a beacon of innovation and perseverance. As the founder of WellNess, an AI-powered health platform, Nyenti has revolutionized access to medical information and services nationwide. Her journey from a curious young girl fascinated by computers to a leading figure in African AI exemplifies the potential of women in STEM when allowed to thrive.

Nyenti's story is one among many inspiring narratives of African women breaking barriers in STEM fields. These trailblazers are making significant contributions to their respective fields and paving the way for future generations of African girls to dream big and pursue careers in science and technology.

Another remarkable example is Jessica O. Matthews, a Nigerian-American inventor and entrepreneur who has made waves in the renewable energy sector. Matthews' invention of the SOCCKET, a soccer ball that generates and stores electricity during play, has brought clean energy to off-grid communities across Africa. Her company, Uncharted Power, continues to innovate in renewable energy and smart infrastructure, addressing critical needs in African communities and beyond.

In South Africa, Ndoni Mcunu, the founder of Black Women in Science (BWIS), has been instrumental in creating a supportive network for African women in STEM. Through mentorship programs, workshops, and advocacy, Mcunu's organization addresses the systemic barriers that often prevent women from pursuing and succeeding in scientific careers (Womentech network, n.d.). These women, and countless others across the continent, are making significant strides in biotechnology and space science. Dr Gladys Kalema-Zikusoka, a Ugandan veterinarian and conservationist, has pioneered innovative approaches to wildlife conservation that integrate human and animal health. In Ethiopia, Betelhem Dessie, often referred to as "Ethiopia's tech queen," began coding at the age of ten and now leads youth programing initiatives at iCog Labs, working on AI projects and inspiring young girls to enter tech fields.

However, the path to success for these women has been far from easy. They have had to overcome numerous challenges that persist in the African STEM landscape. Gender bias remains a significant hurdle, with deeply ingrained societal norms often discouraging girls from pursuing STEM subjects from an early age. In many African cultures, science and technology are still perceived as male-dominated fields, leading to a lack of female role models and mentors (AAUW, n.d.).

The scarcity of resources presents another major challenge. Many African countries struggle with inadequate funding for STEM education and research, a problem that disproportionately affects women who may already be at a disadvantage due to gender biases. Limited access to advanced technologies, well-equipped laboratories, and up-to-date learning materials can hinder the progress of aspiring female scientists and engineers.

Moreover, the lack of supportive policies and institutional frameworks often creates additional barriers. Many African countries have yet to implement comprehensive strategies to promote gender equality in STEM fields, leaving talented women to navigate male-dominated spaces without adequate support or protection from discrimination.

Despite these challenges, African women in STEM continue to persevere and excel. Their achievements are a testament to their resilience, creativity, and determination. Theo Nyenti's WellNess platform has improved healthcare access for thousands of Cameroonians, while Jessica O. Matthews' renewable energy innovations are lighting up communities across the continent. Dr Kalema-Zikusoka's work in conservation has not only protected endangered species but also improved the lives of local communities through eco-tourism and public health initiatives.

These women are not just making strides in their respective fields; they are actively working to create more opportunities for other women in STEM. Many have established mentorship programs, founded organizations to support women in science, or advocated for policy changes to promote gender equality in STEM education and careers (AAUW, n.d.).

The impact of these trailblazers extends far beyond their achievements. They serve as powerful role models, inspiring young African girls to pursue their interests in science and technology. Their success stories challenge stereotypes and demonstrate that women can excel in any field, given the right opportunities and support.

Furthermore, the diverse perspectives and experiences women bring to STEM fields drive innovation in ways particularly relevant to African contexts. From developing mobile applications that address uniquely African challenges to creating sustainable solutions for energy and environmental issues, women in STEM are at the forefront of innovations that are transforming life on the continent.

As we celebrate these achievements, it's crucial to recognize that there is still much work to do to achieve true gender equality in African STEM fields. Increased investment in STEM education for girls, implementing policies to promote women's participation in science and technology, and continued efforts to challenge societal norms and stereotypes are all necessary to create a more inclusive STEM landscape in Africa.

The stories of women like Theo Nyenti, Jessica O. Matthews, and Ndoni Mcunu are potent reminders of what is possible when talent meets opportunity. As we move forward, we must continue supporting, celebrating, and amplifying African women's voices in STEM. Their contributions shape the future of science and technology on the continent and pave the way for a more equitable, innovative, and prosperous Africa for all.

EMERGING TECHNOLOGIES: SHAPING THE AFRICAN CONSUMER LANDSCAPE

The rapid advancement of technology is reshaping the African consumer landscape in profound and far-reaching ways. From the bustling markets of Cairo to the savannah plains of Tanzania, emerging technologies are transforming how people live, work, and interact. Innovations at the forefront of this technological revolution address uniquely African challenges and opportunities, with mobile money and e-agriculture being particularly impactful.

Mobile Money: A Financial Revolution

No other technology has had as significant an impact on the daily lives of African consumers as mobile money. In a continent where traditional banking infrastructure is often limited, especially in rural areas, mobile money platforms have emerged as a game-changer, bringing financial services to millions of previously unbanked persons (Kehinde-Peters, 2024).

The story of mobile money in Africa is inextricably linked to the rise of M-Pesa in Kenya. Launched in 2007 by Safaricom, M-Pesa allows users to deposit, withdraw, and transfer money using mobile phones. The simplicity and accessibility of the system led to its rapid adoption, and it wasn't long before similar services spread across the continent.

According to Jack and Suri (2011), the story of mobile money in Africa is inextricably linked to the rise of M-Pesa in Kenya. Launched in 2007 by Safaricom, M-Pesa allows users to deposit, withdraw, and transfer money using mobile phones. The simplicity and accessibility of the system led to its rapid adoption, and it wasn't long before similar services spread across the continent.

Today, mobile money is used for various transactions, from paying for goods and services to receiving salaries and government benefits. In Nigeria, platforms like Paga and OPay have gained significant traction, offering services beyond simple money transfers, including bill payments, airtime purchases, and savings and loan products.

Mobile money has meant unprecedented convenience and financial inclusion for the average Nigerian consumer. Lagos market traders can now accept digital payments, reducing cash-handling risks. Rural farmers can receive payments for their produce directly on their phones, removing the need for lengthy and potentially dangerous journeys to banks in distant towns (Kremer & Houngbo, 2020).

Moreover, mobile money has opened new opportunities for entrepreneurship and economic growth. Small business owners can now easily track their transactions, manage their finances more effectively, and even access microloans based on their transaction history.

E-Agriculture: Cultivating Success through Technology

Another area where emerging technologies significantly impact is agriculture, which remains crucial to many African economies. E-agriculture, integrating digital technologies into farming practices, is revolutionizing how farmers work and improving agricultural productivity across the continent.

In Nigeria, platforms like Farmcrowdy and Thrive Agric are leveraging technology to connect smallholder farmers with resources, information, and markets. These platforms use mobile apps to provide farmers with real-time weather updates, crop management advice, and market price information. Some even facilitate crowdfunding, allowing urban dwellers to invest in farming projects and share in the profits.

For the Nigerian farmer, e-agriculture tools have become indispensable. A farmer in Kaduna State can now use her smartphone to access expert advice on pest control, optimal planting times, and soil management. She can also connect directly with buyers, cutting out middlemen and securing better prices for her produce.

Drones are another emerging technology making waves in African agriculture. In countries like Rwanda, drones are used for precision agriculture, helping farmers monitor crop health, optimize irrigation, and even deliver supplies to remote areas (Asif, S. et.al. (2023) states). This technology is precious on a continent where many farming regions are difficult to access by traditional means.

The Role of African Women Entrepreneurs

It's important to note that African women entrepreneurs are at the forefront of leveraging these emerging technologies. In the mobile money space, women-led startups are creating innovative solutions tailored to the needs of female consumers.

In the e-agriculture sector, women entrepreneurs are also making significant strides. Peris Bosire, co-founder of FarmDrive in Kenya, developed a mobile phone-based tool that helps small-scale farmers build credit scores and access loans. Her platform has particularly impacted women farmers, who often face more difficulties accessing traditional financial services.

In a male-dominated sector, Anyworkman Services Limited, a women-led, technology-enabled artisan management startup, connects individuals and businesses looking for help around the house and in the workplace with vetted, insured, and verified local handymen, including women.

These women-led innovations are not only driving technological advancement but also promoting gender equality in the tech sector. By creating solutions that address the specific needs of women consumers, they ensure that emerging technologies' benefits are more equitably distributed.

Beyond Mobile Money and E-Agriculture

While mobile money and e-agriculture are standout examples, the impact of emerging technologies on African consumers extends far beyond these sectors. In healthcare, telemedicine platforms are bringing medical expertise to remote areas, while AI-powered diagnostic tools are improving the accuracy and speed of disease detection (Khalifa & Albadawy, 2024).

E-learning platforms are making quality education more accessible, a trend that has accelerated due to the COVID-19 pandemic. Companies like Andela, co-founded by Nigerian entrepreneur Iyinoluwa Aboyeji, are using technology to train and connect African software developers with global opportunities, creating new pathways for career development.

According to Oladele (2024), the rise of e-commerce is another significant trend, with platforms like Jumia (often called the "Amazon of Africa") transforming retail across the continent. These platforms are changing how consumers shop and creating new opportunities for small businesses to reach broader markets.

In the energy sector, pay-as-you-go solar systems bring clean, affordable electricity to off-grid communities. Companies like M-KOPA Solar in East Africa and Lumos in Nigeria use mobile technology to allow consumers to pay for solar power in small, affordable increments, much like they would for mobile airtime.

As we look to the future, emerging technologies such as blockchain, AI, and the Internet of Things (IoT) promise to bring even more transformative changes to the African consumer market. Blockchain technology, for instance, is being explored for applications ranging from land registry systems to supply chain management, potentially increasing transparency and reducing corruption.

The rapid adoption and innovative application of these technologies demonstrate the dynamism of African markets and the ingenuity of African entrepreneurs. However, it's crucial to note that the benefits of these technologies are not yet universally accessible. Issues such as the digital divide, with disparities in access between urban and rural areas and between men and women, remain significant challenges.

Nevertheless, the overall trajectory is one of positive change. Emerging technologies are not just improving consumer convenience but fundamentally altering the economic landscape, creating new jobs, fostering entrepreneurship, and driving financial inclusion. As these technologies continue to evolve and spread, their potential to catalyze sustainable development and improve the quality of life across Africa is immense.

The story of emerging technologies in Africa involves innovation, adaptation, and transformation. It's a narrative driven by African consumers' and entrepreneurs' needs and creativity, with women playing an increasingly central role. As we move forward, ensuring that these technological advancements benefit all segments of society will be crucial in realizing the full potential of Africa's technological revolution.

CONSUMER BENEFITS AND CHALLENGES

The rapid proliferation of emerging technologies across Africa has brought about a change in basic assumptions in the consumer landscape, offering myriad benefits while presenting unique challenges. As we delve into this complex interplay of advantages and obstacles, it's crucial to examine how these technologies are reshaping the lives of African consumers, with a particular focus on women's experiences.

Benefits for Consumers

1. **Increased Access**

One of the most significant benefits of emerging technologies for African consumers is the unprecedented increase in access to services and information. Mobile technology has been a game-changer, bringing the power of the internet and digital services to millions previously excluded.

In Nigeria, for instance, mobile phones have become a gateway to a wide range of services. A woman in a remote village in Enugu State can now access banking services, healthcare information, and educational resources, all from a device that fits in her pocket. This increased access has been particularly transformative for women, who often face more significant barriers to mobility and access to traditional services.

For instance, Dr Ola Orekunrin, a medical doctor and founder of Flying Doctors Nigeria, has revolutionized emergency healthcare access. By leveraging technology to coordinate air ambulance services, she has made critical medical care accessible to people in remote areas of Nigeria. This innovation has saved countless lives and demonstrated how technology can bridge geographical gaps in essential services.

In education, platforms like She Leads Africa, co-founded by Yasmin Belo-Osagie, have made entrepreneurship and career development resources more accessible to young African women. These platforms have democratized access to knowledge and mentorship through online courses, webinars, and networking opportunities, empowering a new generation of female leaders and entrepreneurs.

E-commerce platforms have expanded consumer choice, allowing individuals to purchase products nationwide or internationally. This is especially beneficial for consumers in rural areas who previously had limited access to a diverse range of goods.

2. **Improved Efficiency**

Emerging technologies have significantly improved efficiency in various aspects of consumer life. Mobile money platforms, for example, have streamlined financial transactions, saved consumers time, and reduced the risks associated with carrying cash.

In agriculture, precision farming techniques enabled by drones and IoT sensors help farmers optimize their yields and reduce waste. For a female farmer in Kenya, this could mean more accurate information about when to plant, irrigate, and harvest, leading to improved productivity and income.

Artificial intelligence is driving efficiencies in customer service, with chatbots and virtual assistants providing 24/7 support for common queries. This is particularly beneficial for time-pressed consumers, who can now get instant responses to their questions or concerns.

Companies like Hello Tractor, founded by Jehiel Oliver, have introduced an "Uber for tractors" model in agriculture. This platform connects tractor owners with farmers who need their services, optimizing resource utilization and increasing agricultural productivity. While not founded by a woman, this company exemplifies the kind of efficiency-driven innovations transforming African industries, and similar models are being adopted and adapted by women entrepreneurs across the continent.

3. **Affordability**

Many emerging technologies have made services more affordable for African consumers. The pay-as-you-go model, popularized by mobile telecommunications, has been adapted to other sectors like energy and water, making these essential services more accessible to low-income consumers.

In the solar energy sector, companies like M-KOPA and Lumos allow consumers to pay for solar home systems in small, manageable installments, often lower than what they previously spent on kerosene or generator fuel.

The rise in mobile money has also reduced the cost of financial transactions, particularly for remittances. This has been a boon for many African families who rely on money sent from relatives working in urban areas or abroad.

In healthcare, companies like 54gene, founded by Dr Abasi Ene-Obong and Damilola Oni, are working to make genetic testing more affordable and accessible to African populations. This has the potential to revolutionize personalized medicine and drug development, ultimately leading to more cost-effective treatments for African patients.

POTENTIAL CHALLENGES

While the benefits of technological advancements are significant, it's essential to acknowledge and address the challenges that have emerged alongside these innovations.

1. **Digital Divide**

Despite the increased access to technology, a significant digital divide persists in Nigeria and across Africa. This divide is multifaceted, encompassing disparities in internet access, digital literacy, and device access (Fawowe & Araba, 2024).

According to the Nigerian Communications Commission, as of 2021, about 50% of Nigerians had access to the internet. While this represents significant growth over the past decade, it still means that half of the population is potentially left behind in the digital revolution.

The divide is particularly pronounced along urban-rural lines. Urban areas generally have better infrastructure and higher incomes, making it easier for residents to access and use digital technologies. Rural areas, on the other hand, often lack reliable electricity and internet connectivity, making it challenging for residents to participate in the digital economy.

There's also a gender dimension to the digital divide. Women and girls in Nigeria are less likely to have access to mobile phones and the internet than their male counterparts. This gap is often due to lower incomes, cultural norms, and lower digital literacy rates among women.

To address this challenge, initiatives like the Women's Technology Empowerment Centre (W.TEC), founded by Oreoluwa Somolu Lesi, are working to bridge the gender digital divide. W.TEC provides technology education and mentoring for girls and women, helping to increase their digital literacy and participation in the tech sector.

2. **Data Privacy Concerns**

As more Nigerians adopt digital services online, data privacy and security concerns have become increasingly prominent. Many consumers are worried about how their personal information is collected, used, and protected by tech companies and government agencies.

These concerns are founded. In recent years, several high-profile data breaches in Nigeria have affected both private companies and government institutions. For example, in 2019, a significant data breach at a foremost Nigerian financial institution exposed millions of customers' personal and financial information.

Moreover, there's growing awareness about the potential misuse of personal data for targeted advertising, political manipulation, or surveillance. The Cambridge Analytica scandal, which had implications for Nigeria's 2015 elections, highlighted the potential for data misuse on a large scale.

Addressing these concerns requires a multipronged approach. Nigeria has taken steps to protect consumer data on the regulatory front with the Nigeria Data Protection Regulation (NDPR) of 2019. However, enforcement remains challenging, and many consumers still need to know their rights under this regulation.

Tech companies, particularly those led by women, also play a role in addressing these concerns. For instance, PrivateChefs, a Nigerian startup co-founded by Niniola Soleye, focuses on providing secure, privacy-focused cloud services for businesses. Such initiatives demonstrate how the tech industry itself can lead in prioritizing data privacy and security.

3. **Job Displacement**

The rise of automation and artificial intelligence has led to concerns about job displacement across various sectors. While technological advancements create new job opportunities, they also have the potential to eliminate certain types of jobs, particularly those involving routine tasks.

Manufacturing, customer service, and data entry in Nigeria are particularly vulnerable to automation. For instance, the increasing use of chatbots and AI-powered customer service solutions could reduce the need for human customer service representatives.

However, it's important to note that technology creates new job opportunities. The challenge lies in ensuring the workforce is equipped with the skills needed for these new roles. This is where initiatives focused on upskilling and reskilling become crucial.

Organizations like Andela, co-founded by Iyinoluwa Aboyeji, address this challenge by providing intensive software development training to young Africans. While not exclusively focused on women, Andela's model demonstrates how technology education can create new employment opportunities in the digital economy.

Similarly, initiatives like GirlCode, founded by Zandile Keebine in South Africa but reaching across the continent, focus specifically on equipping girls and women with coding skills. Such programs help address potential job displacement and work toward closing the gender gap in tech.

Navigating the Challenges: The Role of African Women Consumers

As we confront these challenges, African women consumers have a crucial role in shaping the trajectory of technological development on the continent.

Digital Literacy and Empowerment

To bridge the digital divide, African women must actively seek digital literacy and empowerment opportunities. This could involve participating in free online courses, attending local tech workshops, or joining women-focused tech communities.

Initiatives like the African Girls Can Code Initiative, a joint program of the African Union Commission, UN Women, and the International Telecommunication Union, provide a model for large-scale efforts to increase digital literacy among young African women. Consumers can advocate for and support the expansion of such programs in their local communities.

Data Privacy Advocacy

Regarding data privacy, African women consumers can play a pivotal role by demanding transparency from tech companies and holding them accountable for their data practices. This involves reading privacy policies, understanding how personal data is used, and making informed choices about which services to use.

Consumers can also advocate for more robust data protection laws and better enforcement of existing regulations. African women can help create a safer digital environment for all users by raising awareness about data privacy issues and supporting privacy-focused initiatives.

Continuous Learning and Skill Development

Continuous learning and skill development are crucial to navigating the changing job landscape. African women consumers should prioritize lifelong learning, stay abreast of technological trends, and acquire new skills that are in demand in the digital economy.

This could involve taking online courses, participating in coding boot camps, or pursuing further education in STEM fields. Platforms like Coursera, edX, and local initiatives like ALX offer a wealth of resources for skill development.

Supporting Women-Led Tech Initiatives

By consciously choosing to support tech products and services developed by African women, consumers can help drive the growth of women-led businesses in the tech sector. This contributes to a more inclusive tech landscape and ensures that the solutions being developed are more likely to address the unique needs and perspectives of women consumers.

Engaging in Tech Policy Discussions

Finally, African women consumers can discuss tech policy, participate in public consultations, and make their voices heard on issues that affect them. This could involve joining local tech meetups, participating in online forums, or engaging with policymakers through social media and other platforms.

By actively participating in these discussions, African women can help ensure that policies and regulations in the tech sector consider their needs and perspectives.

In conclusion, while the technological revolution brings both benefits and challenges for Nigerian consumers, it also presents an opportunity for African women to actively shape the future of technology on the continent. By embracing digital literacy, advocating for data privacy, continuously developing their skills, supporting women-led initiatives, and engaging in policy discussions, African women consumers can play a crucial role in creating a more inclusive and equitable tech landscape.

THE FUTURE: A MORE INCLUSIVE TECH LANDSCAPE

As we look toward the future of technology in Africa, it's clear that fostering a more inclusive tech landscape is not just a matter of equality—it's a crucial factor in driving innovation, economic growth, and societal progress. The participation of African women in STEM fields is central to this vision, and there are several key areas where focused efforts can make a significant difference.

Encouraging Girls' Interest in STEM Education

The journey toward a more inclusive tech landscape begins with education. Encouraging girls' interest in STEM early on is crucial to building a pipeline of future women leaders in technology.

Early Exposure and Engagement

Research has consistently shown that early exposure to STEM can spark lifelong interest and passion (Stofan, 2019). In Nigeria and across Africa, there's a growing recognition of the need to introduce girls to STEM concepts and role models at a young age.

Initiatives like the Lagos State Eko-Konnect Girls in STEM program are leading the way. This program organizes workshops, mentorship sessions, and hands-on activities for girls in secondary schools, introducing them to various aspects of STEM in an engaging and relatable manner.

Similarly, the African Women in Science and Engineering (AWSE) organization runs outreach programs in schools across several African countries, including Nigeria. These programs expose girls to STEM concepts and introduce them to successful women in STEM fields, providing crucial role models.

Addressing Stereotypes and Biases

One of the significant barriers to girls' participation in STEM is the persistence of stereotypes and biases that portray these fields as masculine domains. Addressing these misconceptions is crucial to creating a more inclusive environment.

Educators, parents, and community leaders are vital in challenging these stereotypes. This involves encouraging girls to pursue STEM subjects and actively countering messages that might discourage them.

Media representation is another powerful tool in this regard. Initiatives like "Hidden Figures: The Nigerian Edition," a project that highlights the stories of Nigerian women in STEM, help provide visible role models and counter prevailing stereotypes.

Curriculum Development and Teaching Methods

How STEM is taught can significantly impact students' interest and engagement. There's a growing recognition of the need to develop curricula and teaching methods that are more inclusive and appealing to diverse learners.

In Nigeria, organizations like the Science Teachers Association of Nigeria (STAN) are working to promote more engaging and hands-on approaches to science education. These methods often involve project-based learning and real-world applications, which can be particularly effective in capturing and maintaining students' interest.

Additionally, incorporating examples and case studies that feature women scientists and engineers can help girls see themselves represented in these fields. The "Heritage and Legacy of Women in Science for Development" project, initiated by UNESCO, provides resources for educators to integrate the stories of African women scientists into their teaching.

Initiatives to Support and Empower African Women in STEM

While encouraging girls' interest in STEM is crucial, supporting and empowering women already pursuing or working in STEM fields is equally important. Several initiatives across Africa are working toward this goal.

Mentorship Programs

Mentorship plays a crucial role in supporting women in STEM fields. Programs that connect established professionals with young women entering the field can provide guidance, support, and valuable networking opportunities.

In Nigeria, the Women in Technology in Nigeria (WITIN) organization runs a mentorship program that pairs young women in tech with experienced professionals. This program provides career guidance and helps participants navigate the unique challenges that women often face in male-dominated tech environments.

Similarly, the TechWomen program, while not exclusively African, has provided valuable mentorship opportunities for many Nigerian women in tech. This U.S. Department of State initiative connects emerging women leaders in STEM from Africa, Central and South Asia, and the Middle East with their counterparts in the United States for a mentorship and exchange program (TechWomen, n.d.).

Networking and Community Building

Creating spaces for women in STEM to connect, share experiences, and support each other is another crucial aspect of empowerment. These communities can provide a sense of belonging and mutual support in often isolating environments.

The Nigerian Women in Tech (NWIT) network is an excellent example of such an initiative. NWIT organizes regular meetups, conferences, and online discussions, creating a vibrant community for women in tech across Nigeria.

On a broader scale, the African Women in Technology (AWIT) conference, held in various African countries, including Nigeria, provides a platform for women in tech to network, share knowledge, and inspire each other.

Funding and Resources

Access to funding and resources is often a significant barrier for women in STEM, particularly for those looking to start their own tech ventures. Addressing this gap is crucial for empowering more women to take leadership roles in the tech sector.

Initiatives like She Leads Africa's accelerator program provide funding, mentorship, and resources specifically for women-led startups in Africa. Similarly, the African Women Innovation and Entrepreneurship Forum (AWIEF) runs programs that provide funding and support for women entrepreneurs, including those in tech fields.

Policy Advocacy

Empowering women in STEM also involves advocating for policies that promote gender equality in these fields. This includes policies related to education, workplace discrimination, parental leave, and funding allocation.

Organizations like the Association of Professional Women Engineers of Nigeria (APWEN) are crucial in advocating for policies supporting women's participation and advancement in engineering (APWEN, n.d.).

The Positive Impact of Inclusivity on Technological Innovation

As we work toward a more inclusive tech landscape, it's essential to recognize that this isn't just about equality—it's about unleashing the full potential of Africa's talent pool and driving innovation that benefits all of society.

Diverse Perspectives Drive Innovation

According to Okatta et. al. (2024), research has consistently shown that diverse teams are more innovative. When people from different backgrounds and experiences come together, they bring a more comprehensive range of perspectives to problem-solving, leading to more creative and comprehensive solutions.

In the African context, including more women in tech teams means that African women's unique experiences and perspectives are more likely to be considered when developing new technologies. This can lead to innovations that better serve the needs of all members of society.

For instance, the success of M-Pesa, the mobile money transfer service that has revolutionized financial inclusion in Kenya and beyond, was partly attributed to its appeal to women users. The service was designed with input from diverse teams, including women, which helped ensure it addressed the needs of female consumers who were often underserved by traditional banking systems.

The Role of NGOs in Promoting Women in STEM: A Nigerian Woman's Journey

Women remain a minority in the tech industry, constituting only 28% of professionals globally and 30% in Sub-Saharan Africa. This disparity, as portrayed in the statistics, shows the need for gender equality in STEM fields (UNESCO in UNDP, 2019). A picture was posted on the "streets of Instagram" with the caption, "You will know she is studying a serious course when most of her course mates are male." It was a picture of course mates after their six-month Industrial Training (IT) defense and presentation, which engineering students usually undergo. With one female and about 12 young men, hence the caption.

The comments under this post, even from the female folks, were all in sync with the caption and further buttressed the under-representation of women in STEM until today, 2024. This is even more so in Africa, with its age-old belief that a woman's education ends in her husband's kitchen; hence, parents seem to prioritize the education of their male children over that of the female, let alone encourage her to veer into STEM. The gender disparity in the STEM profession emerges from the stereotype that men are better suited for it than women, leading to a subconscious agreement in the girl child, who ends up settling for other fields out of fear of failure and social acceptance, stifling and limiting the girl child from even trying to take her rightful place in the STEM profession.

The Great Potential That Lies Within Women

The question that begs for an answer is whether male folk have a characteristic trait that seems to make them a better fit for STEM professions, or if it is all a case of societal prejudice. For the few who dare, statistics show that women have all it takes to break the glass ceiling and rise above societal norms and beliefs in the classroom and beyond. Reflecting on their journey as STEM learners, parents are often ready to go above and beyond to ensure their children, male and female alike, are given the wings to fly in their chosen field of endeavor.

The girl child, saddled with the responsibility of having to prove herself much more than her male counterparts, is spurred to put her very best into her studies and excel in STEM subjects, as seen in her outstanding success, usually as the best graduating student in her chosen STEM courses amid male course mates who are in the majority. Female STEM learners are known to graduate with exceptional grades and have gone ahead to do extraordinary things in the tech space. From the foregoing, it is safe to say that females and males alike have all it takes to make their marks as STEM professionals.

The capacity of women in the transformation and development of a nation cannot be overemphasized, leading to a growing recognition of the impact of women in STEM in solving life's challenges and leading innovations that can transform lives and contribute to sustainable development.

According to Tunga (n.d.), some of the outstanding African women in tech who have made significant contributions to the tech ecosystem, empowering others and driving innovations in their respective fields, to mention but a few, are as follows: Odunayo Eweniyi (Nigeria) co-founded Piggyvest, a popular savings and investment platform; Jihan Abass (Kenya) launched Lami Insurance Technology, a digital platform; Ire Aderinokun (Nigeria) co-founded Helicarrier, a cryptocurrency infrastructure company; Abiola Rasaq (Nigeria) founded The Bug Detective, a community for software testers; Ada Nduka (Nigeria) founded SheCode Africa, an organization promoting women in tech; and Rebecca Enonchong (Cameroon) founded Appstech and is a prominent advocate for technology in Africa.

Balancing Family and Career

As much as women triumph in STEM careers, firsthand experiences abound of the challenges women face on this journey, such as marriage to an unsupportive spouse, childbirth, balancing family and career, and finding employment in a competitive job market. It takes sheer doggedness to rise above these obstacles, remain committed to one's career goals, and pursue opportunities to gain practical experience.

Such is the case of a female computer science graduate who married an unsupportive partner, had to take a detour into education, and accepted a secondary school ICT teacher position. While the teaching job may seem like a detour from a tech career and aspirations, it provided her with valuable teaching experience and an in-depth understanding of the importance of education in fostering technological literacy and encouraging girls to pursue the STEM career pathway.

Dasgupta and Stout (2014) aptly state that traditional teaching methods often fail to captivate girls in STEM education due to their preference for practical and experiential learning. Hence, there is a need for alternative 21st-century teaching and learning methods, such as guided inquiry, problem-based learning, and project-based learning, which empower learners to build their understanding through exploration and experimentation. This makes STEM learning more engaging and fosters female learner's confidence and interest in these fields.

With the teaching career also came the appreciation of the fact that there is no one-size-fits-all approach to teaching and learning. Hence, there is a need to accommodate the needs of diverse learners through scaffolding—offering support to students as they grapple with new concepts and differentiated instructions, which entail tailoring learning activities and resources to meet the unique needs of individual students, and the Universal Design for Learning (UDL) framework. All aim to craft accessible and engaging learning environments, ensuring all learners, regardless of their individual needs, can thrive and achieve their full potential in the ICT classroom so that no child is left behind.

Rediscovering Passion and Embracing Opportunities

Dr Elizabeth Jumoke Eterigho, the immediate past president of the Association of Professional Women Engineers of Nigeria, in her interview with Funmi Ogundare of ThisDay, quoted the World Economic Forum's Global Gender Gap Report of 2017 as reporting that ***"the female talent remains one of the most under-utilized business resources, either squandered through lack of progression or untapped from the onset."*** (Eterigho, 2022).

From this statement, the case of our case study may be seen as one of lack of progression, as she felt stuck in the classroom with a craving for opportunities to contribute to the STEM profession more meaningfully. She sought opportunities to improve and expand her knowledge toward a career in data analysis. During this period, she discovered the invaluable support of NGOs and tech training programs targeted at reducing the gender disparity in the tech profession.

During one of her numerous online searches for free tech training, she discovered a gold mine, Tech4Dev (Technology for Social Change and Development Initiative). Tech4Dev is a nonprofit social enterprise dedicated to empowering Africans with digital skills and creating opportunities for decent work. Tech4Dev focuses on marginalized communities and strives to create a more equitable society through digital empowerment and advocacy. Under the leadership of Oladiwura Oladepo, their Women Techsters initiative aims to train 5 million women across Africa by 2030, bridging the digital gender gap and ensuring equal access to technology opportunities (TECH$DEV, 2021).

It is so fulfilling to speak to the girls in public schools in one's community and have their eyes light up at emerging technologies such as Generative Artificial Intelligence (ChatGPT); it is such a wholesome experience to be the harbinger of good news that their STEM dreams are valid.

The Impact of NGO Initiatives on Women in Tech and Alignment with the SDGs (Sustainable Development Goals)

NGOs are vital in promoting women's participation in STEM fields in Africa. They provide accessible and affordable educational opportunities, establish STEM centers, advocate for gender equality, offer mentorship and networking opportunities, and support women-led startups. These efforts are crucial in addressing the challenges faced by women in STEM and creating a more inclusive and equitable environment for all. Some of these NGOs include the Consumer Advocacy and Empowerment Foundation (CADEF), Tech4Dev Women Techsters Fellowship, CodeVixens, Africa Agility Foundation, Girls in Tech (GIT), and the GetBundi Tech Sis program.

All these programs provide the participants with the necessary tech training, mentorship, and networking opportunities to develop their skills, build a strong portfolio, transition, and get jobs in the tech space. Through the support of these programs, women have successfully transitioned into tech careers, leveraging their newfound expertise to contribute to various projects and initiatives.

By focusing on education, gender equality, economic growth, innovation, and sustainable development, NGOs contribute significantly to a more just and equitable society. These efforts align with the Sustainable Development Goals (SDGs) of ensuring inclusive education, empowering women and girls, promoting sustainable economic growth, building resilient infrastructure, and fostering innovation (United Nations Department of Economic and Social Affairs, Sustainable Development, n.d.).

CONCLUSION

The case study's tech journey highlights some of the challenges women face in STEM, the power of perseverance and determination, and NGOs' crucial roles in helping women reach their full potential and make their mark in tech. In addition to the extraordinary efforts and giant strides made by NGOs in promoting gender equality in education and STEM fields in Africa, there is a need for the government, well-meaning citizens, and all stakeholders to provide women with the necessary financial support through bursaries, financial aid, and scholarships to give them the soft landing they deserve and the push to follow their dreams, create supportive work environments, offer mentorship and networking opportunities, and raise awareness and orientation/reorientation for the girl child.

REFERENCES

AAUW. (n.d.). The STEM Gap: Women and Girls in Science, Technology, Engineering and Mathematics. American Association of University Women. https://www.aauw.org/resources/research/the-stem-gap/

APWEN. (n.d.). Who we are? The Association of Professional Women Engineers of Nigeria APWEN. https://www.apwen.org.ng/about

Asif, S. et.al. (2023). Drones as Precision Farming Tools: Enhancing Efficiency and Sustainability. Research Gate. https://www.researchgate.net/publication/373977177_Drones_as_Precision_Farming_Tools_Enhancing_Efficiency_and_Sustainability

AUDA-NEPAD. (September 15, 2022). Heightening The Participation of African Women in Science, Technology, Engineering, and Mathematics Career Paths. African Union Development Agency. https://www.nepad.org/blog/heightening-participation-of-african-women-science-technology-engineering-and-mathematics

Dasgupta, N., & Stout, J. G. (2014). Girls and Women in Science, Technology, Engineering, and Mathematics: STEMing the Tide and Broadening Participation in STEM Careers. *Policy Insights from the Behavioral and Brain Sciences*, *1*(1), 21–29. http://journals.sagepub.com/doi/pdf/10.1177/2372732214549471

Eterigho. (January 12, 2022). Eterigho: Women's Participation in STEM Key to 2030 Agenda for SDGs. https://www.thisdaylive.com/index.php/2022/01/12/eterigho-womens-participation-instem-key-to-2030-agenda-for-sdgs/

Fawowe, S. & Araba, F. (2024). Bridging the digital divide: Exploring the Impact of Modern Technology on Education for Nigerian Children. *International Journal of Advanced Research in Multidisciplinary Studies (IJARMS)*, *4*(1). ISSN 2756-4444. E-ISSN 2756-4452. https://alhikmahuniversity.edu.ng/centralJournal/my_portal/user/event/bookUrl914.pdf

Jack, W. & Suri, T. (2011). Mobile Money: The Economics of M-PESA. National Bureau of Economic Research. https://www.nber.org/system/files/working_papers/w16721/w16721.pdf

Kehinde-Peters, O. (August 31, 2024). Fintech: A Pathway to Closing Africa's Gender Gap in Financial Inclusion. Nigerian Tribune. https://tribuneonlineng.com/fintech-a-pathway-to-closing-africas-gender-gap-in-financial-inclusion/

Khalifa, M. & Albadawy, M. (2024). AI in Diagnostic Imaging: Revolutionising Accuracy and Efficiency. *ScienceDirect*, *5*, 100146. https://www.sciencedirect.com/science/article/pii/S2666990024000132

Kremer, M. & Houngbo, G. (July 9, 2020). *Grow Back Better? Here's How Digital Agriculture Could Revolutionise Rural Communities Affected by COVID*. World Economic Forum. https://www.weforum.org/stories/2020/07/digital-agriculture-technology/

Okatta, C. & Ajayi, F. & Olawale, O. (2024). Enhancing Organizational Performance Through Diversity and Inclusion Initiatives: A Meta-Analysis. *International Journal of Applied Research in Social Sciences*, *6*, 734–758. https://www.researchgate.net/publication/380115625_ENHANCING_ORGANIZATIONAL_PERFORMANCE_THROUGH_DIVERSITY_AND_INCLUSION_INITIATIVES_A_META-ANALYSIS

Oladele, E. (2024). E-Commerce Boom in Africa: Driving Growth and Transforming Retail in 2024. 1st Afrika. https://1stafrika.com/2024/09/12/e-commerce-boom-in-africa-driving-growth-and-transforming-retail-in-2024/

Stofan, E. (2019). An Early Start in STEM Can Spark a Lifelong Passion. *Smithsonian Air & Space Magazine*. https://www.smithsonianmag.com/air-space-magazine/early-start-stem-180973139/

TECH4DEV. (2021). *Empowering Communities in Africa with Skills for Decent Work Using Technology to Advance Sustainable Human Capital Development in Africa*. TECH4DEV Brochure. Technology for Social Change and Development Initiative. https://tech4dev.com/assets/images/brochure/Tech4DevBrochure.pdf

TechWomen. (n.d.). What Is TechWomen? https://www.techwomen.org/

Tunga. (n.d.). *50 Outstanding African Women in Tech*. African IT Talent Startup. https://tunga.io/50-outstanding-africanwomen-in-tech/

UNESCO in UNDP. (January 31, 2019). *Creating a Level Playing Field for Women in Technology in Africa*. United Nations Development Program. https://www.undp.org/africa/press-releases/creating-level-playing-field-women-technology-africa

United Nations Department of Economic and Social Affairs Sustainable Development (n.d.) Transforming Our World: the 2030 Agenda for Sustainable Development. https://sdgs.un.org/2030agenda

Womentech network. (n.d.). How Can Mentorship Programs Break Barriers for Women in Leadership Roles? https://www.womentech.net/en-za/how-to/how-can-mentorship-programs-break-barriers-women-in-leadership-roles

8 The Future of Retail Banking and Financial Services in Africa and the Role of Women in STEM in Advancing the Growth and Transformation of the Sector

Ngozi Megwa

SECTION ONE

THE FUTURE OF RETAIL BANKING

A combination of macroeconomic and geopolitical factors, including technology-driven innovation and industry disruptive forces will continue to change the shape of the retail banking sector globally.

A wide range of disruptive forces will continue to shape the future of retail banking globally. Whilst ongoing macroeconomic factors, geopolitical shocks and regulatory uncertainties create a mixed outlook for CEOs, the sector will continue to grapple with other disruptive forces as customer preferences advance, markets fragment, technology drives innovation and new business models evolve. This dynamic evolution in the sector means that retail banking globally will continue to face a more complex environment, exacerbated by growing competition from digital banks, including declining customer loyalty and attrition.

Advances in technology are driving a new wave of innovation in the retail banking sector in what is frequently referred to as the digital age. Therefore, thriving in the future will hinge on agile, forward-thinking business strategies centred on sustainable growth. Retail banks that rethink customer experiences, explore innovative ways to deliver value, drive revenue through business model innovation, whilst improving efficiency, will set themselves apart, ensuring success whilst others may fall behind.

 DOI: 10.1201/9781003540618-8

Personalization will become critical towards maintaining customer loyalty and increasing customer lifetime value. However, challenges with legacy banking infrastructure and technology, including limited access to real-time customer data and predictive insights – along with concerns about data privacy, fraud and risk, may continue to hinder banks' efforts to craft truly personalized customer experiences.

To remain competitive in this new digital era, retail banks will need to find ways of integrating digital technologies and solutions throughout their value chain, ensuring these efforts are both sustainable and focused on tangible outcomes. Banks will require the ability to modernize their technology infrastructure and capabilities in a way that allows them to embrace advanced modelling tools leveraging artificial intelligence and machine learning to enable the generation of predictive insights for customers. Enhanced enterprise risk management processes are paramount, providing the needed flexibility and agility to adopt evolving business models.

Globally, the retail banking sector should prioritize the following five key areas – enhancing customer engagement and loyalty, leveraging data access, analytics and insights, building robust cybersecurity and fraud prevention processes, embracing technology-led digital transformation and continuing to implement innovative digital payment solutions across various customer segments.

We are already starting to see this manifest in how banks are investing. About 80% of all financial institutions plan to increase technology spend over the next two years, with fraud detection/mitigation, digital banking and data analytics expected to be the top three technology investments in 2024 and 2025.[1]

The banking sector's spending on generative artificial intelligence (AI) is projected to surge to 84.99 billion US dollars by 2030, with a remarkable 55.55% compound annual growth rate. This astonishing growth reflects the increasing significance of AI in reshaping the banking industry's landscape, as it seeks to leverage AI-driven technologies to enhance customer experiences, optimize operations and drive innovation.[2]

SECTION TWO

Retail Banking in Africa – 'The Future'

The African banking sector continues to show resilience and growth since the pandemic.

A McKinsey report highlights that the pandemic's impact on African banks in 2020 was milder than predicted, as government data and financial indicators from key economies such as Kenya, Morocco, Nigeria and South Africa revealed less disruption than initially anticipated. In Africa, the banking sector is anticipated to exhibit a steady growth rate, with net interest income expected to grow at a compound annual growth rate (CAGR) of 1.22% between 2024 and 2029, resulting in a market volume of US $244.20 billion by 2029.[3]

Whilst these trends hold true for banking across the continent, retail banking does constitute a significant portion of Africa's banking industry. In 2017, retail banking revenues were approximately $35 billion, accounting for about 40.7% of the continent's total banking revenues of $86 billion.[4]

It is assumed that the retail banking sector across Africa has continued to grow and adapt in the post-pandemic era driven by factors including but not limited to increased financial inclusion, technology-driven digital banking innovations and economic resilience. As of 2022, Africa's retail banking sector was projected to reach $53 billion in revenues, accounting for approximately 41% of the continent's total banking revenues of $129 billion – thus a significant driver of growth and sustainability for the African banking sector.

In summary, whilst African retail banks have demonstrated resilience and recovery since 2022 and share several post-pandemic trends with their global counterparts, digital transformation, advanced customer centricity, data intelligence and enhanced cybersecurity, they also exhibit unique adaptations, particularly in building resilience across key banking performance indicators, which could impact their ability to compete effectively in the digital era.

Banking Performance Indicators (BPIs)

Cost-to-Income Ratio (C/I): African banks generally exhibit higher cost-to-income (C/I) ratios compared to their global counterparts. According to McKinsey, banks in most African markets have high cost-to-income (C/I) ratios along with low banking penetration compared with benchmark markets in the Middle East, Eastern Europe, Southeast Asia and Latin America. Additionally, they are costly to run with an average cost-to-asset ratio of between 4 and 5%, almost twice as high as the global average. In 2021, the average C/I ratio across 133 countries was approximately 54.8%. Notably, Germany reported a high ratio of 88.62%, whilst China maintained a low ratio of 31.75%. In contrast, African banks tend to have higher C/I ratios. For example, Nigeria reported a ratio of 65.04% in 2021.[5]

The ongoing efforts by banks across Africa to improve their cost-to-income ratios are showing positive results. In the first quarter of 2024, some of Nigeria's tier-one banks reported declines in their C/I ratios, indicating improved cost management amid rising inflation. For instance, United Bank for Africa (UBA) recorded a C/I ratio of 57.8%, down from 61.1% in the same period of 2023.[6]

According to PwC's analysis, South Africa's major banks had a combined C/I ratio of 52.2% in March 2024, showing an improvement from 53% in the previous year. This suggests efforts to enhance operational efficiency. However, many still face challenges in achieving operational efficiency comparable to global standards.

This disparity, compared with global averages, could be attributable to factors including but not limited to higher costs in technology and infrastructure including regulatory compliance. With these costs typically rising faster than revenue growth, it underscores the need for enhanced operational efficiencies. Addressing these challenges is crucial for improving profitability and competitiveness in the global banking sector.

Return on Equity (ROE)

The picture for African banks is mixed, with some banks matching and, in certain instances, exceeding global averages. However, the overall trend suggests room for improvement in profitability and operational efficiency to consistently align with global standards. Recent statistics in 2022 highlight that, since 2016, the average ROE in Africa's five largest banking markets has declined by approximately 2.6 percentage points due to decreasing net interest income and persistent high operating costs.[7]

Additional headwinds include intensified competition from new banks and fintechs, particularly in the provision of digital banking services.

To address these profitability challenges will require the banks to look at opportunities to increase productivity across the banking value chain in an agile and sustainable way.

SECTION THREE

Retail African Banking in Africa – Growth Levers

To compete effectively in the dynamic and rapidly evolving financial landscape across Africa, retail banks need to adopt a multifaceted strategy that embraces technology-driven innovation (digital transformation) with a focus on advanced digital and personalized customer experiences whilst maintaining robust cybersecurity and compliance frameworks.

A continued focus on financial inclusion, leveraging public and private partnerships including fintechs to drive an agile, resilient and sustainable financial services ecosystem is also important. This will help them thrive in the digital era and remain competitive in a market characterized by rapid change and growing customer expectations.

Almost all African banking groups say that in 2024 and beyond, digital transformation is a top priority. According to a report by the African Business, at least half (51%) of the 153 African banks surveyed in the African Digital Banking Transformation Report 2023 consider digital transformation to be the one of the most important factors in their growth strategies. In total, 95.5% have it among their top three priorities, citing investments in digital transformation as a key imperative. Pan-African banking groups, or banks that operate in several African countries, face a more complex and demanding digital transformation journey than their local counterparts. They must deal with challenges that include language and regulatory differences, diverse technology ecosystems and customer profiles and high costs of cross-border integration.

Transformative Progress in the Retail Banking Sector Will Be Powered By Key Growth Enablers

Advanced Digital Experiences and Personalization

Prioritizing customer engagement and loyalty is an important strategy for African retail banks to achieve profitable growth. Competitive dynamics in the retail banking sector have created a battleground for the consumer's share of wallet, which means that banks must find new ways of developing deeper customer relationships.

Personalization and customer empowerment are becoming critical to demonstrating customer lifetime value. Getting this strategy right can generate sustainable growth in the form of higher deposits and higher profitability as customers avail of more banking products and services with lower volatility. This creates a more resilient and stable revenue stream from loyal customers.

To do this effectively, banks will be required to become adept at using customer data to curate products and services to the 'power of one', including the delivery of real-time financial advice and solutions powered by advanced modelling tools that generate predictive insights.

The risks of not getting personalization right can have implications both on the customer experience overall and negatively impact the financial performance and operational resilience of the bank.

In the context of banking, this focus on customer centricity allows banks to forge deeper and more relevant relationships with their customers, which can only be made possible by getting the concept of personalization right. This will require them to be able to customize the right products and services through tailored and curated experiences. This can only be made possible with access to the right customer data and advanced modelling tools that generate predictive insights and analytics. The goal is to leverage insights to inform real-time decision-making, which can only be made possible by having modular and composable platforms (ERP, CRM, Digital Banking) and an agile technology stack to manage the end-to-end consumer experience and service delivery, irrespective of the customer's point of interaction with the bank. In other words, an integrated technology and banking engagement platform.

This requires an effective customer engagement ecosystem embedded across key components of the retail banking value chain, encompassing the following key building blocks:

Multi-Channel Consistency

Retail banks will need to ensure that they are able to provide a consistent experience across all channels, whether online, in-store, in-app or in branch to ensure that customers receive a uniform service and messaging to enhance trust and satisfaction. Today, the majority of platforms that exist across the banks don't operate across all the bank's channels; therefore, customers are not able to switch between channels in a consistent and seamless fashion. The challenge for the banks is that they are not able to get a single consolidated view of the customer interactions across all channels to allow for a consistent and personalized experience. This leads to banks being unable to access and share data across all channels in real time.

A Robust Data Strategy

Retail banks are consistently looking for innovative ways to strengthen customer relationships and enhance financial empowerment. Personalization will play a crucial role in showcasing long-term value, with data and technology serving as the essential building blocks. A robust data strategy and the right technology infrastructure are essential to understand customer behaviours, preferences and needs.

Advanced analytics and insights are instrumental in predicting trends, personalizing interactions and making informed business decisions that all enhance the overall customer experience.

Data access, storage and quality within the confines of the current banking legacy infrastructure remain a challenge to banks across Africa and globally. Additionally, banks struggle with their ability to get a 360-degree view of customer data across all banking channels. Most of the customer engagement platforms that exist today struggle to integrate various banking functions such as CRM, data analytics, marketing automation and customer feedback. They often don't have the right predictive models or data insights that give real-time insights into customer behaviour across their lifecycle and whole of banking relationship. Data privacy and regulatory considerations are also an important factor

Advanced modelling tools, predictive insights data analytics platforms and digital banking engagement platforms are crucial for success and can solve the challenges banks are facing. Solving these challenges is a key driver of most banks' digital transformation initiatives.

Successful applications of digital transformation in this area have led to improved decision-making, risk management and product innovation. Additional opportunities lie in developing more sophisticated data platforms and employing regulatory technologies (RegTech) to ensure compliance and enhance data security.

Personalized Micro Products and Services

As the retail banking sector in Africa and globally continues to face intensified and sustained pressure from a new wave of 'digital' players in the form of digital-first banks and neo banks including large digital-only players such as telecommunications and data companies, they will have to continually find new and innovative ways to disrupt and reshape the traditional retail banking value chain with innovative solutions.

The telecommunication companies have played a very strong role in the growth and diversification of mobile financial services delivered to different segments of the African population through the mobile money ecosystem. This has and will continue to advance financial and digital inclusion across the African continent. The growth of mobile financial services in Africa by these players is because of their intentional and deliberate ecosystem strategy of partnering with a myriad of fintech's and other third-party providers, operationalizing and embedding optionality of products and services into their offerings to end consumers.

To stay competitive, banks must prioritize product and service innovation, embracing and adopting new roles within the value chain, powered by agile methodologies. They need the agility to deconstruct the value chain based on real-time, personalized customer needs and then reconfigure it in novel ways to unlock diverse growth opportunities.

The real challenge banks are grappling with is their ability to curate, deliver and communicate these new value propositions for customers across the whole of the bank, based on personalized customer needs, including the ability to identify upsell and cross-sell opportunities in the process in an agile and timely manner.

Integrated Technology and Engagement Platform

The traditional banking model is built on legacy systems that are costly to maintain and restrict the banks' ability to innovate with the agility required in today's digital age. This is particularly true in the African banking sector, where productivity remains a challenge for the sector. Traditional banking technology falls short of supporting the digital transformation objectives essential in today's financial services sector, particularly as banks strive to find new ways to deliver value, boost revenue and enhance efficiency. Banks are not able to seamlessly orchestrate customer and employee journeys across all banking channels, nor are able to connect channel activity to the banks' downstream systems of records in real-time.

There's been a major paradigm shift in the banking industry towards a new operating model that's centred around customer engagement orchestration and the bank's ability to create differentiating value propositions. Digital banking engagement platforms that are composable, modular and can interface seamlessly with legacy banking, including downstream banking applications, can be a viable solution as part of a bank's digital transformation agenda.

Optimization and Feedback Mechanisms

The ability to establish clear metrics to measure the effectiveness of customer engagement platforms, programmes and initiatives is particularly important given the competitive dynamics at play in the financial services industry today. The ability to leverage these metrics and emerging technology (such as AI) regularly and systematically gather and analyze feedback across channels and the customer journey, including assessing progress and areas for improvement, is crucial towards ensuring an active and engaged customer.

IMPACT OF STEM ON CUSTOMER CENTRICITY IN RETAIL BANKING AND FINANCIAL SERVICES

Data Analysis and Customer Insights

STEM skills in the areas of advanced data modelling, analysis and statistics can help banks in Africa to better understand customer behaviour, preferences and buying patterns. The ability to analyze, understand and predict customer behaviour supports personalization efforts with the associated benefits of higher satisfaction, loyalty, retention and sustainable growth.

Predictive Analytics and Personalization

Predictive analytics, made possible by STEM disciplines in the areas of data science, statistics and machine learning, enables businesses to anticipate customer needs. This positively impacts both customer loyalty and engagement by offering customers products and services that are curated and relevant to them in real time and proactively builds trust and reinforces loyalty and drives engagement.

To be sure, STEM advancements, particularly in artificial intelligence (AI) and machine learning (ML), are revolutionizing customer engagement by enabling personalized banking experiences, advanced data modelling and predictive analytics and insights. Successful implementations include AI-powered chatbots and personalized marketing strategies that enhance customer satisfaction and loyalty. Opportunities exist in further harnessing AI to anticipate customer needs and provide tailored services. However, gaps remain in achieving seamless integration across all customer interaction channels.

Technology and Automation Platforms

Customer engagement platforms such as CRM systems, chatbots, AI-driven analytics and automated service systems leverage STEM skills in the engineering and computer science disciplines to advance and power seamless customer interactions.

This has a positive impact on driving customer loyalty as these platforms power more efficient services, faster response time and efficient issue resolution and feedback loops – i.e., consistent and reliable quality of support. These automated systems also support loyalty programmes and personalized rewards, fostering long-term customer relationships.

AI and machine learning enable brands to offer dynamic experiences such as personalized content, targeted recommendations and interactive features, keeping customers engaged with the brand.

Customer Feedback and Sentiment Analysis

Natural language processing (NLP) and machine learning (ML), part of the STEM skills, enable companies to perform sentiment analysis on customer feedback and reviews. By collating and understanding customer sentiments and feedback in real-time, banks in Africa can quickly address issues or concerns, improving satisfaction and building loyalty. This is especially crucial as banks embark on the digital transformation journey in the digital age. From a customer engagement standpoint, engaging customers through surveys, feedback forms and real-time responses demonstrates that a company values their opinion, increasing their willingness to engage with the banks and building more trust.

Innovation and Product Development

Innovations driven by STEM professionals in product development (for example, creating more efficient or smarter products through engineering or technology) contribute to improved product offerings. Impact on Loyalty: Companies that innovate and consistently improve their products or services are more likely to retain customers, as they offer superior value compared to competitors. Impact on Engagement: Engaging customers with cutting-edge products, apps or solutions that are continuously improved keeps them invested and eager to participate in beta tests, reviews or brand communities.

To summarize, STEM disciplines can contribute significantly towards enabling and developing **Customer Centricity** across the banks in Africa as they embark on the next click of transformation in the digital age. Integrating these STEM-driven strategies into banking operations and the broader banking value chain supports the creation of more personalized, efficient and seamless customer experiences, fostering engagement and long-term loyalty.

Cybersecurity and Fraud Management

As African economies continue to promote and support inclusive digital financial services (DFS) growth across both the public and private sectors, it is imperative to continually enhance cybersecurity, financial crime and overall enterprise risk management to mitigate the ever-increasing threats. According to the Global Findex database, Africa leads the world in mobile money adoption, which is one of the primary drivers of digital financial inclusion. With over one billion transactions processed per month in Africa, the addressable market that needs to be protected is sizeable.

According to an article by the CyberFI project, between January and July 2023, the top 3 sectors that experienced the most attacks were the financial services sector, followed by telecommunications companies and government agencies. According to the report, Africa reportedly loses about 4 billion dollars per annum to cybercrime.

As digital banking expands, robust cybersecurity measures and enterprise risk management become increasingly crucial. Successful implementations include advanced encryption, biometric authentication and AI-driven threat detection systems that protect sensitive data and ensure transaction integrity. Opportunities lie in enhancing cybersecurity frameworks and proactive risk management practices to address evolving threats. Continuous updates and addressing sophisticated cyber threats are essential gaps that need to be bridged.

Impact of STEM on Cybersecurity and Enterprise Risk Management in Retail Banking and Financial Services

The correlation between STEM and Cybersecurity and Enterprise Risk Management in financial services is significant. The STEM disciplines, by their very nature, provide the foundational skills, knowledge and specialism needed to understand, address and mitigate the threats.

Technical Knowhow (Infrastructure Development, Network Security)

STEM skills in computer science, engineering (especially systems and network engineering) and mathematics are fundamental in developing, securing and maintaining the relevant IT networks and systems. Understanding how systems and networks operate supports the creation of robust risk mitigation strategies. Engineers assess potential points of failure or failure of vulnerability in systems and design risk controls to limit exposure.

Network segmentation, intrusion detection systems and secure architecture are some examples of how STEM plays a direct role in fortifying cybersecurity, with built-in security mechanisms to prevent unauthorized access, malware attacks and data breaches.

Predictive Analytics, Risk Modelling

Data science, statistics and machine learning play a key role in analysing large datasets to predict and prevent cybersecurity threats. Mathematics and statistics are heavily used in developing models to quantify cyber risks and predict the likelihood and impact of various risks. STEM-based risk models help organizations quantify and assess risks. Predictive analytics can forecast potential cyberattacks or system failures, enabling companies to proactively manage risks by strengthening weak points in their defences.

Incident Response, Forensics, Cryptography

Professionals with strong STEM backgrounds (particularly computer science, IT and mathematics) can use forensics tools to interrogate and understand the various threats, such as tracing the origins of a cyber-attack and recovering lost or compromised data.

Cryptography ensures the confidentiality, integrity and authenticity of data. Mathematics, particularly fields like theory, algebra and probability form the backbone of cryptography, which is essential for securing digital communications.

To summarize, in Africa, where financial systems are still undergoing a period of growth and transformation, fraud, security breaches and cyber risks are a real threat to the ecosystem.

STEM-driven cyber and risk management solutions such as encryption, secure authentication methods including biometric and identity solutions, intrusion detection systems and machine learning algorithms can help banks in Africa protect sensitive financial data, omni-channel digital transaction and critical infrastructure, leading ro a more resilient financial services ecosystem in Africa.

Digital Payments Innovation

Digital payments innovation continues to transform Africa's banking and financial services landscape, driven by mobile money dominance, with Africa as the global leader in mobile money adoption and mobile financial services, the rise of fintechs who are revolutionizing digital payments by offering fast, secure, innovative solutions for businesses and individuals as well as the rapid growth of e-commerce and cross-border payments.

Digital payments innovation in Africa is reshaping the continent's financial services landscape, with innovative products and solutions leapfrogging that of more developed economies. With the combination mobile of financial services, fintech solutions and blockchain backed by government initiatives, Africa continues to power digital and financial inclusion across the continent.

IMPACT OF STEM ON DIGITAL PAYMENTS INNOVATION IN RETAIL BANKING AND FINANCIAL SERVICES

AI and Machine Learning for Personalized Financial Products

Personalization is key to customer lifetime value, loyalty and retention. AI-powered financial services driven by STEM skills in data science and mathematics can provide tailored lending, investment and insurance products better suited to individual needs. This can benefit small businesses, farmers and individuals by offering them access to financial products that they might not have been able to obtain from traditional banks, especially when they lack formal financial histories or credit scores.

Expanding Access to Credit and Microfinancing

There is still a population of Africans (consumers and small businesses) across multiple industry sectors who lack access to financial services. STEM-based disciplines such as data science and analytics can support financial institutions to better assess creditworthiness by leveraging alternative data sources such as mobile phone usage, social media behaviour and transaction history.

With STEM-based analytics, financial institutions can leverage alternative data from third-party providers to assess the credit risk of these individuals, opening access to credit for millions. This can boost entrepreneurship and economic development by providing small businesses and individuals with the necessary capital to grow. Companies like Branch, Tala, Jumo and Fido are leveraging alternative data points such as mobile phone usage, transaction histories and behaviours data to assess creditworthiness and provide loans to people who lack traditional financial histories.

Blockchain, Smart Contracts and Decentralized Finance (DeFi)

Blockchain can revolutionize banking by enabling secure and efficient cross-border payments, reducing transaction fees and preventing fraud (create linkage to MA?). It also has the potential to create more transparent financial transactions, which is especially important in regions with limited trust. Blockchain technology relies heavily on STEM disciplines like cryptography and software engineering and can provide secure, transparent and tamper-proof ledgers for banking transactions. Countries like Nigeria, South Africa, Kenya and Ghana are leading blockchain deployment in Africa in the financial service sector. Blockchain technology is increasingly viewed as a tool to improve transparency, enhance financial inclusion and solve structural challenges within the sector

STEM disciplines can also drive the development of decentralized finance (DeFi) applications and smart contracts, which allow financial transactions to be executed automatically based on pre-programmed conditions for use cases such as peer-to-peer lending, decentralized insurance and automated savings programmes.

Workforce Augmentation and Digital Skills

As digital transformation takes hold within the financial services sector globally, automation and artificial intelligence are replacing human thinking and urging banks to revisit their talent landscape and the skills required to stay ahead of the curve. Additionally, trends such as the gig economy and crowdsourcing are changing the way work is done. Given the profitability conundrum of banks in Africa, it's important to think through how they can successfully integrate the digital and human workforce, including how they will change their operating models to remain relevant.

Automation, AI and robotics are increasingly seen as tools to transform traditional banking workflows, promoting enterprise agility and adaptability.

IMPACT OF STEM ON ENTERPRISE AGILITY AND THE FUTURE OF WORK IN RETAIL BANKING AND FINANCIAL SERVICES

Automation and Robotics for Operational Efficiency

Robotics and automation, supported by engineering and computer science, can streamline banking operations and reduce manual processes.

Automation tools like Robotic Process Automation (RPA) can help banks process routine tasks such as compliance checks, account opening and loan processing faster and more efficiently. This reduces human error, speeds up service delivery and cuts operational costs, freeing up human resources for more complex, value-added tasks.

STEM Education and Workforce Development

Fostering STEM education across Africa is crucial for developing a skilled workforce that can drive technological innovation in the banking sector.

By promoting STEM education, banks can develop local talent capable of designing and maintaining digital banking platforms, cybersecurity systems and data analytics solutions. This is essential for long-term sustainability, as local talent is more likely to develop solutions tailored to Africa's unique challenges.

Numerous use cases are emerging to achieve efficiencies, accelerate productivity and reimagine customer experiences. Gen AI can help improve essential banking functions by taking on many data-heavy tasks. Internally, banks can unlock hidden insights, forecast with better precision and augment staff productivity.

SECTION FOUR

The Impact of Women in STEM Roles in Advancing the Growth and Transformation of the Retail Banking Sector in Africa

The retail banking and financial services industry in Africa is on an extraordinary journey of digital transformation, leveraging technology and innovation to rethink the customer experience, business models and operating models. The goal is to look for agile and innovative new ways to deliver value, generate revenue and improve efficiency with a focus on customer lifetime value. This transformation has created a new demand for knowledge and experience in the field of science, technology, engineering and maths (STEM) and highlights the unique role women in STEM can play in spearheading and accelerating growth and transformation in the sector.

Globally, technology adoption is tipped to remain a key driver of business transformation in the next five years, according to the Future of Jobs report[1]

Over 85% of organizations surveyed identify increased adoption of new and frontier technologies and broadening digital access as the trends most likely to drive transformation in their organization. More than 75% of companies are looking to adopt these technologies in the next five years. From an African perspective, the same story holds true. Over 51% of a sample size of 153 African banks surveyed in an African Digital Banking Transformation Report consider digital transformation to be one of the most important factors in their growth strategies, with over 95% naming it among their top three priorities. Empowering women in these fields can lead to a more robust, equitable and dynamic retail banking sector and brings with it unique benefits ranging from digital transformation and innovation in fintech, product innovation founded on rich diverse perspectives, to driving inclusive and sustainable growth which ultimately increases the competitiveness of the sector within the global retail banking and financial services market.

Digital Transformation and Innovation in Fintech

Women in STEM are already leading the charge in creating innovative fintech solutions that enhance digital banking services across the continent. Leveraging and adopting emerging technologies, they are leading innovation in areas such as artificial intelligence (AI) and machine learning to automate processes, improve fraud detection and personalizing customer experiences. Women often excel in designing user-friendly interfaces and customer experiences that cater to diverse user needs. Their understanding of the challenges faced by underbanked populations allows them to create solutions that are accessible and easy to use.

Some examples of women-led fintechs in the sector include:

YAPILI: Co-founded by Linda M. Mungai, YAPILI is a digital platform that connects health practitioners and patients, integrating fintech solutions to improve healthcare financing and access to medical services.

Lidya: Co-founded by Tunde Kehinde and Ebi Atawodi, Lidya is a fintech company that provides loans to small and medium-sized enterprises (SMEs) in Nigeria and other African countries. It focuses on helping businesses access the capital they need to grow.

Kudi: Founded by a group of women, Kudi is a Nigerian fintech company that provides a platform for payments and financial services, particularly aimed at underserved populations.

Benevolent: Co-founded by Hilda M. Sulaiman, Benevolent is a social enterprise platform that connects people willing to help with individuals in need of financial support, facilitating small loans for personal projects.

Bringing Innovation and Diverse Perspectives

Women in STEM roles in financial services bring unique perspectives and approaches to problem solving, leading to more creative and innovative solutions in designing banking products and services that are inclusive and cater to underserved segments such as women entrepreneurs (SMEs) and rural communities. This diversity of thought can lead to creative ideas in areas such as fintech and digital banking. They are able to identify with and relate differently to these segments, thereby providing relevant insights that cater to the specific needs and behaviours of these end consumers.

Driving Inclusive and Sustainable Growth

Access to digital financial services (DFS) is crucial in addressing many of Africa's socio-economic challenges as well as supporting the closure of the gender gap in STEM careers. Women in STEM roles within banking and financial services can play a key role in championing and design inclusive financial products that could support in addressing this inequality. The IFC found that women-owned businesses in Africa are underfunded, with only about 7% of women entrepreneurs receiving funding compared to 16% of their male counterparts. Women-owned businesses receive only about 3% of total bank lending in Africa, indicating a significant funding gap.

The role model concept also works effectively. Having women in STEM roles in financial services can serve as an encouragement for more women to enter STEM roles, ultimately contributing to the broader goals of gender equality and workforce diversification, both of which are key to sustainable economic development.

Increasing Sector Competitiveness in the Global Market

Given the industry-disruptive forces in the banking and financial services sector, workforce agility is necessary to adapt to fast-changing global technological trends. Having a balanced gender workforce in STEM roles can give African banks a competitive edge in the global market. Firms that demonstrate diversity in leadership and can demonstrate innovation agility and business model innovation, supported by strong female representation in STEM roles, are likely to appeal to investors looking for diverse, forward-thinking organizations.

Harnessing the power of Networks and Strategic Partnerships

Women STEM networks focused on retail banking and financial services in Africa are essential for promoting gender equity and empowering women in these fields. By providing mentorship, networking opportunities and professional development, these networks help women navigate challenges and contribute to the growth and transformation of the financial sector on the continent.

Mentorship and Support Networks: Women-led fintech initiatives often emphasize mentorship, creating networks that support aspiring female entrepreneurs in tech and finance. Initiatives like the African Women in Fintech & Payments (AWFP) network are working to connect and empower women in the industry. Others include She Leads Africa (SLA), Women in Banking and Finance (WIBF), Women in Finance Network (WIFN) and a host of others playing a crucial role in the space.

Collaboration with Other Enterprises: Many women-led fintech companies collaborate with traditional banks, NGOs and other stakeholders to create holistic financial solutions that can better serve underserved populations.

Building Alliances: Women in STEM can create and participate in networks that connect banking professionals, technologists and entrepreneurs. These networks facilitate knowledge sharing, collaboration and partnership opportunities that can drive innovation and engagement and act as a catalyst to attract more interest and talent to STEM roles in the financial services sector.

Public-Private Partnerships: They can play a key role in fostering partnerships between the private sector and government agencies, aimed at addressing systemic challenges in the banking sector.

Promoting the Sustainability and Ethical Banking Agenda

As awareness of climate change grows, African banks are increasingly adopting sustainable finance practices. By 2025, many banks will integrate environmental, social and governance (ESG) criteria into their lending decisions and investments.

Impact: This shift towards green banking will attract global investors looking to support projects that contribute to sustainable development goals (SDGs), further driving Africa's economic growth.[9]

Women in STEM roles within financial services and banking are often advocates for sustainable and ethical banking practices. Their involvement can drive the development of green finance products and support initiatives that prioritize social and environmental responsibility. They also tend to lead and influence the CSR strategies of banks, ensuring that they contribute positively to the communities they serve, thereby enhancing their reputation and customer loyalty.

SECTION FIVE

Summary

To summarize, it's important to highlight that overall, women in STEM within financial services and retail banking in Africa face greater challenges and lower representation compared to their counterparts in other regions.

From a workforce composition standpoint, data from various studies and reports show that men typically occupy a larger percentage of roles in financial services and banking.

Women constitute about 30% of the workforce in financial institutions across all of Africa, highlighting a substantial gender gap in employment.[1]

For instance, a 2021 report by the World Bank found that women hold only about 20% of senior management positions in the financial sector across many African countries, creating a gender imbalance in leadership. In a study of the banking sector in South Africa, only 25% of executive committee members were women, with an even lower percentage in senior management. Additionally, women occupy only 16% of board positions in African banks, illustrating a critical lack of female representation in governance.[10]

To be sure, whilst STEM has the potential to transform the sector, the extent of its impact will depend on addressing several critical factors, including but not limited to a lack of technological infrastructure, the high cost of technology implementation, regulatory and compliance challenges and cultural and behavioural barriers.

Despite these barriers, there is significant potential for growth and improvement if targeted initiatives are implemented to support women's participation and advancement in these fields. Increasing access to education, mentorship and leadership opportunities for women in Africa's financial services sector could lead to enhanced performance and innovation, ultimately benefiting the broader economy.

Women in STEM talent in Africa are already playing a key role in transforming the growth of retail banking and financial services in the African continent. It's important that we continue to encourage the growth of talent and equip more women to capitalize on the new range of opportunities in the banking and financial services industry as technology-driven innovation and business growth in the sector create opportunities to position Africa as a leading player in global markets.

REFERENCES

1. Jack Henry's 2024–2025 Strategy Benchmark study.
2. bankautomationnews.com, 2024.
3. africanexponent.com, 2025.
4. McKinsey – "Roaring to life: Growth and innovation in African retail banking,", 2018.
5. tradingeconomics.com/Nigeria.
6. businessday.ng "How Nigerian big banks performed in cost-to-income ratio', July 2024.
7. McKinsey, African Banking, The Productivity Opportunity, 2022.
8. Fair Issac Report, March 2023.
9. KPMG, ESG in Practice, results from a global bank survey, 2024.
10. CWDI 2015 Report on Africa (Africa Development Bank).

9 Empowering Change
The Role of Women in STEM for Achieving Sustainable Development Goals

Khawaja Sazzad Ali, Quazi Tafsirul Islam, and Md. Shamim Talukder

INTRODUCTION

The Sustainable Development Goals (SDGs) are a set of 17 global objectives adopted by all United Nations (UN) member states in 2015, aimed at addressing a range of social, economic, and environmental challenges by 2030 (Allen et al., 2018). They were established to sustain the accomplishments of the Millennium Development Goals (MDGs) that have been in place from 2000 to 2015, and their mission is aimed at eradicating extreme poverty, hunger, and diseases (Sachs, 2012). The roots of the SDGs can be traced to the 1992 Earth Summit in Rio de Janeiro, which laid the groundwork for global cooperation on sustainable development. This agenda was later reiterated at the Rio+20 conference in 2012, formulating the current SDGs. The SDGs have since become the central guiding framework for global development, aiming to create a more equitable and sustainable world.

The SDGs were adopted as part of the 2030 Agenda for Sustainable Development at the United Nations Sustainable Development Summit in New York. The SDGs were developed through a participatory process involving extensive consultations with various stakeholders, including governments, civil society, and the private sector (Mishra et al., 2024). This culminated in the adoption of the 2030 Agenda by all UN member states in September 2015, which emphasized that no one should be left behind in pursuing sustainable development. Unlike the MDGs, which primarily targeted developing nations, the SDGs apply universally to all countries, recognizing the interconnectedness of development issues (Sachs, 2012). The 17 goals cover many areas, including poverty eradication, gender equality, climate action, economic growth, and global partnerships.

Global Significance and Targets for 2030

The Sustainable Development Goals (SDGs) are a global framework to address humanity's most critical challenges. Building on the successes and limitations of the

DOI: 10.1201/9781003540618-9

Millennium Development Goals (MDGs), the SDGs adopt a more comprehensive and inclusive approach to development (Stewart, 2015). The goals are formulated so that none of them can be achieved separately from the other, thus emphasizing the need for a coordinated approach to sustainable development (Stafford-Smith et al., 2017). The 2030 targets aim to guarantee that no one is left behind in adopting and practicing inclusive and sustainable economic development, social development, and environmental protection (Schmidt-Traub et al., 2017). The achievement of the SDGs is tracked through 169 targets, with 248 indicators used to assess measures and programs in one or several countries or regions (Mishra et al., 2024). All these measures are instrumental in tracking the advancement toward achieving the SDG's goals at the international level.

There have been some good results in areas such as poverty eradication, improved education, and clean energy, but the 2030 goals are still a long shot. Globalization's challenges, such as the COVID-19 virus outbreak, inter-geopolitical unrest, and environmental pollution, have stalled many advancements, most significantly in health, equity, and climate change (Henderson & Loreau, 2023). As of 2024, significant acceleration is needed across many SDGs to meet the 2030 targets, especially in combating hunger, improving gender equality, and tackling environmental sustainability challenges.

Importance of Women in STEM for Achieving SDGs

Science, Technology, Engineering, and Mathematics (STEM) are important in sustaining innovation and solving world challenges; thus, STEM is needed to support the achievement of the SDGs. Technology and scientific discoveries play a crucial role in the enhancement of growth in fields such as health, agriculture, and energy, which are essential to development (Fonseca et al., 2020). STEM education disciplines enable one to solve problems ranging from climate change to the management of resources and sustainable urbanization (Crespo et al., 2017).

The involvement of women in STEM is critical to accomplishing the SDGs, especially in education, economic growth, gender equality, and innovation. Goals 4 and 5 prioritize quality education and gender equality, and advancing women in STEM minimizes inequities and improves problem-solving from varied viewpoints. Female participation in STEM is critical for economic growth and productivity (Goal 8), as diverse workforces are more creative. Countries that increase women's representation can unlock untapped potential, encourage equitable technological development (Goal 9), and ensure that innovations are inclusive (Kemechian et al., 2023).

Furthermore, empowering women in STEM has a ripple effect, inspiring younger generations of girls to pursue these fields, creating a cycle of sustained growth and innovation. Hence, achieving gender parity in STEM is fundamental for the holistic realization of the SDGs by 2030.

Gender Equality as a Cornerstone for Progress in STEM

Gender equality in STEM is a social justice issue and a strategic imperative for meeting the SDGs. Women account for half of the worldwide population. Therefore, their underrepresentation in STEM disciplines represents a substantial loss of human potential, impeding innovation, economic development, and social improvement (Richardson et al., 2020; Schmader, 2023). Gender equality in STEM is critical for unlocking the full potential of human ability, which supports inclusive growth and development.

Diverse viewpoints in STEM research and problem solutions are critical to addressing complex global concerns. Women provide distinct perspectives and experiences, which boost creativity and result in more complete, equitable, and prosperous solutions (Biermann et al., 2017). This is especially crucial when dealing with climate change, health crises, and technological advancement because multiple methods produce superior results.

Gender equality in STEM contributes directly to SDG 5 (Gender Equality), but its impact is far-reaching. It supports SDG 8 (Decent Work and Economic Growth) by encouraging innovation and entrepreneurship and SDG 10 (Reduced Inequalities) by supporting equal opportunities. Furthermore, women's inclusion in STEM accelerates progress toward SDG 9 (Industry, Innovation, and Infrastructure) by pushing technical growth and ensuring that innovations benefit all sectors of society (Sen, 2019).

Contribution of Women in STEM to Economic Growth and Innovation

Including women in STEM fields significantly contributes to economic growth and innovation. Women in STEM drive technological advancements and are instrumental in developing new products, processes, and services to improve quality of life and promote sustainable development (Stewart-Williams & Halsey, 2021). Studies have shown that companies with higher gender diversity in their workforce tend to perform better financially and are more innovative (Grainger-Brown & Malekpour, 2019). Furthermore, women's participation in STEM can help address skills shortages and meet the growing demand for STEM professionals in various industries (Ike et al., 2019).

The involvement of women in STEM fields is essential for innovation and economic success. According to Pham and Triantis (2015), STEM employment is expected to grow faster than non-STEM jobs, with over 6.6 million positions expected to be filled by 2022. Despite making up almost half of the American workforce, women are still underrepresented in STEM, holding fewer than 25% of these positions (Beede et al., 2011). Up to 2% more GDP might be generated globally if more women pursue careers in STEM fields (Fatourou et al., 2019). There is less of a gender wage gap because women in STEM disciplines make 33% more money than women in non-STEM industries (Beede et al., 2011).

Furthermore, STEM entrepreneurs express greater happiness with their jobs (Blume-Kohout & Adhikari, 2016). Gender stereotypes, a lack of role models, and a lack of family-friendly regulations are obstacles that prevent women from participating (Beede et al., 2011). It is imperative to tackle these concerns to promote creativity, boost the economy, and encourage upcoming generations of females in STEM fields (Fatourou et al., 2019).

The Sustainable Development Goals provide a comprehensive framework for addressing global challenges and achieving a sustainable future. Including women in STEM is essential for leveraging the full potential of innovation and driving progress toward these goals. By promoting gender equality in STEM, we can enhance economic growth, foster innovation, and ensure that sustainable development benefits all members of society.

THE ROLE OF WOMEN IN STEM IN AFRICA

Women in Africa constitute approximately 30% of STEM researchers, which reflects a significant gender disparity in these fields. This underrepresentation is more pronounced as one moves up the academic and professional ladder (Fisher et al., 2020). While women enroll in STEM programs at the undergraduate level, their numbers dwindle significantly in postgraduate and professional roles. According to a study of African women in STEM PhD programs, women tend to publish fewer papers and take longer to complete their PhDs than their male counterparts (Fisher et al., 2020).

In addition, several African countries' organizational cultures still have difficulty ensuring women occupy senior positions (Babalola et al., 2021). Nevertheless, African women engaged in STEM leadership are equally required to practice work-life balance, establish goals, solve challenges, and embrace diversity (Babalola et al., 2021). Currently, undergraduate females in STEM areas can do mentored training through programs such as the African Research Academies for Women (ARA-W) summer research program, which seeks to tackle these disparities (Commodore-Mensah et al., 2020). Of them, 65 independent research projects conducted during the last three years are malaria transmission dynamics, in-vitro cytotoxicity evaluation, high-performance liquid chromatography, palliative care training for doctors, and plastic waste recycling into biodiesel. The young women who participated in the study were of an average age of 20 years; 76% of them studied science, while 11% studied technology, and 14% studied engineering (Commodore-Mensah et al., 2020).

Scholarship, supportive organizational structures, dedication, and hard work all lead to women's STEM success in Africa (Babalola et al., 2021). It is critical to address interlocking systems of oppression and offer targeted support and mentorship opportunities to improve African women's representation and retention in STEM (Commodore-Mensah et al., 2020; Idahosa & Mkhize, 2021).

Historical Context and Progress over Time

Women in Africa have occupied crucial positions throughout history and participated in agricultural, trade, and educational fields. Sudarkasa (1986) noted in his research that African women had significant roles in leadership and decision-making in their communities before colonization. Nevertheless, when colonialism came and other changes in the sociopolitical aspect, women were sidelined, especially in education and professional fields. However, there have been some achievements in women's rights and advancements in the twenty-first century, but the low uptake of women in STEM areas across Africa owing to several factors such as resource constraints, gender biases, and lack of professional development among teachers is a matter of concern (Kiernan et al., 2023). Some of the barriers observed are racism, sexism, as well as the combination of race, gender, and class, which makes it even harder for African women in STEM to advance and stay in their STEM careers (Idahosa & Mkhize, 2021).

In the mid-1990s, several international and regional reforms were initiated to enhance the role of women in STEM education. For example, the Beijing Platform for Action 1995 and the Durban Declaration 2002 stressed the importance of equal access to education and decision-making (Ogunsanya, 2007). They have included estimates of women in higher education and other science-related technology engineering and mathematics courses. Others, such as the African Research Academies for Women (ARA-W), have also been useful in offering mentorship and research opportunities for aspiring female undergraduates pursuing STEM-related careers, thus promoting the growth of the next generation of women scientists and engineers (Commodore-Mensah et al., 2020).

Key Challenges and Barriers Faced by African Women in STEM Fields

Despite progress, African women in STEM continue to face numerous challenges. These barriers can be broadly categorized into sociocultural, institutional, economic, and personal factors.

Sociocultural Factors

Cultural norms and gender stereotypes often discourage women from pursuing careers in STEM. Women in leadership roles in STEM frequently encounter resistance and are less accepted than their male counterparts (Babalola et al., 2021). The dual burden of professional responsibilities and domestic duties disproportionately affects women, particularly those who are married or have children. This often results in lower productivity and a longer time to degree completion for female PhD students (Fisher et al., 2020).

Institutional Factors

Many academic and research institutions in Africa lack gender-friendly policies supporting women's STEM career progression. Policies such as maternity leave, childcare facilities, and flexible working hours are often inadequate or nonexistent (Babalola et al., 2021). The scarcity of female mentors and role models in STEM fields hampers the professional development of young women scientists. Initiatives like ARA-W aim to address this gap by providing structured mentorship programs (Commodore-Mensah et al., 2020).

Economic Factors

Securing funding for research and professional development is a significant challenge for women in STEM. Women often have less access to financial resources and funding opportunities than their male peers (Panigrahi & Bhandigadi, 2020). Educational institutions in many African countries face resource constraints that affect the quality of STEM education. Limited access to laboratories, equipment, and digital technologies further hinders the participation of women in STEM (Mutsvangwa & Zezekwa, 2021).

Personal Factors

Women in STEM often experience lower self-efficacy and confidence in their abilities, affecting their performance and career aspirations. Encouraging environments and positive reinforcement are essential for boosting their confidence (Sikhosana et al., 2023).

The current state of women in STEM in Africa reflects significant underrepresentation and myriad challenges. However, historical progress and contemporary initiatives provide a hopeful outlook for bridging the gender gap in STEM fields. Addressing the sociocultural, institutional, economic, and personal barriers that hinder participation in STEM is crucial for fostering a more inclusive and innovative scientific community in Africa. Efforts such as mentorship programs, supportive policies, and increased funding are vital to empowering African women in STEM and leveraging their potential to drive sustainable development across the continent.

SDGS AND WOMEN IN STEM

SDG 4: Quality Education

SDG 4 focuses on a specific objective of inclusive and equitable quality education and promoting lifelong learning opportunities for all in the context of global sustainable development by 2030. Thus, SDG 4 emphasizes quality education for all people, irrespective of gender, economic status, or if they are part of the marginalized section of society. Priority areas encompass increasing enrollment and completion of primary and secondary education, translating quality learning into reality, early childhood care and education, and developing the requisite skills for post-2015 development, including technological, vocational, and thinking skills. SDG 4 is trying to enable individuals and societies to achieve their potential through education.

African women in STEM hold unique positions in contributing toward the achievement of SDG 4 as much as they fight for access to education with an emphasis on STEM disciplines. They assist in eradicating prejudices and promoting young girls to apply for STEM courses, establishing equity and epoch-making in Africa's educational sector and positively impacting the economy's growth in the continent. They also support mentoring and outreach initiatives that encourage the next generation. The Tanzania Astronomical Cutting Edge Program focuses on girls and women to increase their participation in astronomy and science (Batinoluho, 2024). The African Network of Women in Astronomy and STEM for GIRLS in Ethiopia aims to strengthen girls' and women's involvement in astronomy and science (Pović, 2019).

In Uganda, attempts have been made to eradicate policy weaknesses and improve the stamen education outcomes of female students (Okwakol et al., 2022). The African Research Academies for Women (ARA-W) summer research program offers fellowships and apprenticeships to increase the number of women in STEM careers in Africa (Commodore-Mensah et al., 2020). WITED seeks to improve the intake of female candidates in TVET in Africa, striving to reverse the gaps in education and employment as espoused in the United Nations' Sustainable Development Goals (SDGs) (Mutiku & Kiaritha, 2022). Additionally, there are few women in STEM-related careers in Kenya today due to self-perceptions, teacher/parental expectations, role models and stereotypes working environment, and weak performance (Amunga, 2020).

The challenge of providing quality education is deeply entrenched in the African education system and is closely linked to the broader economic well-being of African nations. While primary school enrollment rates are relatively high, the significantly lower completion rates for secondary education indicate that many students drop out to engage in labor activities (Figure 9.1). According to the 2022 SDG Report, most African countries have primary school enrollment rates exceeding 70%, with South Sudan lagging at 36.98%. Countries such as Seychelles, Togo, Sierra Leone, and Namibia approach nearly 100% enrollment. However, secondary school completion rates present a contrasting scenario. For example, Central African Republic, Equatorial Guinea, Niger, and South Sudan have completion rates of 10.97%, 20.09%, 15.98%, and 9.86%, respectively. Notably, Mauritius leads with the highest secondary school completion rate, while Namibia, Seychelles, and South Africa also show strong performance in this area.

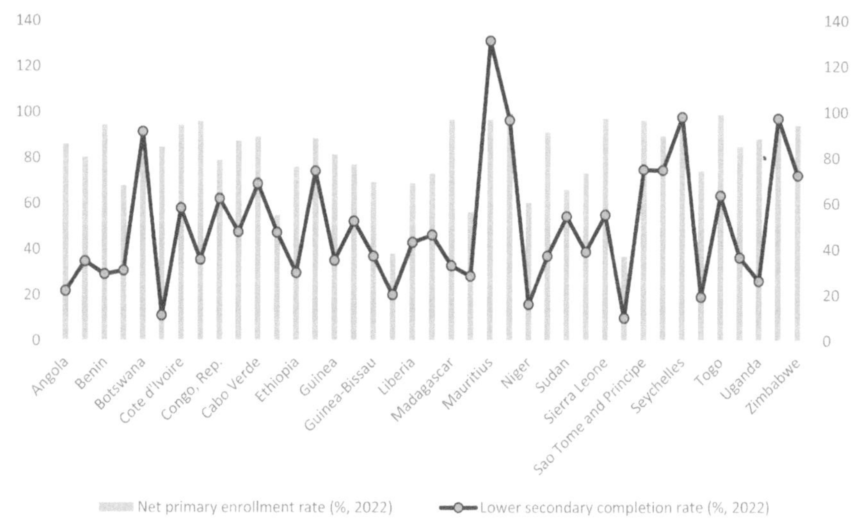

FIGURE 9.1 Primary enrollment and secondary completion rate in African countries in 2022 Author's illustration based on UNESCO data

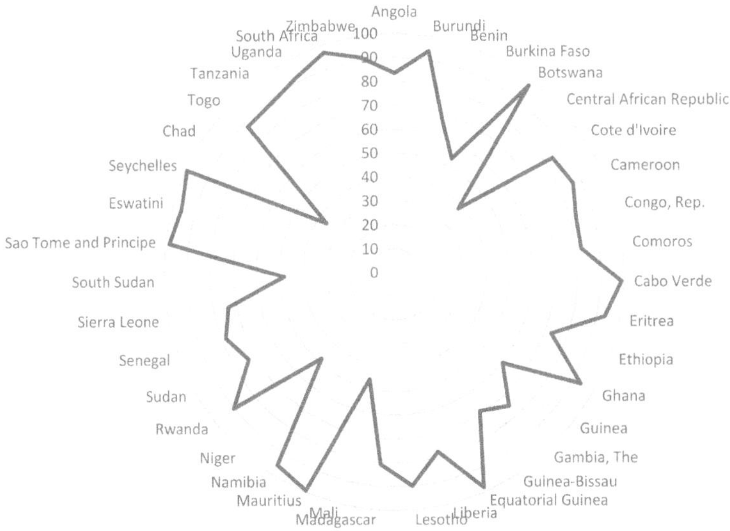

FIGURE 9.2 Literacy rate of African countries (% of population aged 15 to 24) in 2022 Author's illustration based on UNESCO data

This concerning outlook underscores the long-term challenges faced by African nations in achieving universal literacy, as no country in Africa had reached a 100% youth literacy rate as of 2022 (Figure 9.2). Mauritius and Seychelles lead the continent with literacy rates of 99.3% and 99.1%, respectively. However, countries such as Chad and the Central African Republic, with literacy rates of 36.07% and 38.34%, respectively, remain at the lower end of the spectrum. These figures highlight the urgent need for targeted interventions and policies aimed at improving literacy rates, particularly in the most underdeveloped regions. Without concerted efforts, disparities in education are likely to persist, hampering broader socioeconomic development across the continent.

BOX 1: STEMI MAKERS AFRICA—EMPOWERING THE NEXT GENERATION IN STEM

STEMi Makers Africa is a non-governmental organization founded by Amanda Obidike to revolutionize STEM (Science et al.) education across Africa. The initiative seeks to bridge the gap between traditional educational methods and the skills required in today's workforce. By focusing on innovative teaching techniques and project-based learning, the organization empowers educators and students, with a strong emphasis on sparking interest in STEM careers among young women. This approach directly supports SDG 4 by promoting inclusive, quality education for all.

Targeting underserved populations, particularly women and girls, STEMi Makers Africa provides mentorship and training to encourage inclusivity in STEM fields. The organization connects education with real-world applications, preparing young Africans for careers in technology and innovation. Through strategic partnerships with organizations such as Oracle Academy and Microsoft, STEMi Makers Africa has extended its global reach and impact.

Numerous success stories have emerged from the organization's initiatives, demonstrating its transformative impact. Notable examples include:

- **Project Kuongoza Mentoring Program:** Operating in 21 African countries, this program empowers women and girls by providing mentorship and skill development in STEM. Participants have reported increased confidence and a greater ability to pursue careers previously deemed unattainable.
- **Educator Training:** STEMi Makers Africa has trained over 6,500 educators in modern, project-based teaching methodologies. This training has improved student engagement and understanding of STEM subjects, with many teachers reporting enhanced effectiveness in the classroom.
- **STEMi Clubs:** Across 19 countries, 371 STEMi Clubs have been established, offering students hands-on opportunities to explore STEM subjects. Success stories include innovative student projects addressing local challenges like clean water access and renewable energy solutions.

Founder Amanda Obidike's journey, from frustration in her early career to establishing a platform that empowers young Africans, mirrors the transformation experienced by many participants. Her leadership inspires the next generation to acquire the skills and confidence necessary to thrive in tech-related fields.

Sources: HundrED.org; STEMi Makers of Africa; The Nasdaq Entrepreneurial Center

SDG 5: Gender Equality

SDG 5 (Gender Equality) aims to achieve gender equality and empower women and girls by eliminating discrimination, violence, and harmful practices such as child marriage and female genital mutilation. It emphasizes equal opportunities in leadership, education, employment, and resource access. SDG 5 also promotes the formulation of policies and legislation that foster gender equity, ensuring women's full participation in social, political, and economic spheres. Attaining SDG 5 is vital for establishing inclusive societies where women and men enjoy equal rights and opportunities.

African women in STEM play a pivotal role in advancing SDG 5 by challenging entrenched gender biases in male-dominated fields and promoting women's empowerment. Their presence in STEM helps dismantle stereotypes and barriers that deter other women from entering these disciplines. Their success highlights the importance of inclusive policies and practices that support gender parity. Mentorship and networking opportunities are essential for fostering empowerment and creating supportive environments for women in STEM (Konyeha et al., 2021). Through their achievements in STEM, these women serve as role models advocating for equal opportunities in education and the workforce. Their participation in STEM influences policies aimed at narrowing the gender gap, driving socioeconomic development, and ensuring women play an equal role in shaping Africa's future innovations.

Despite financial challenges, institutions like the Women's University in Africa contribute to SDG 5 by increasing female enrollment and offering empowerment programs (Nondo & Mbereko, 2020). Additionally, Science Granting Councils (SGCs) in Africa are crucial in promoting gender equality in Science, Technology, and Innovation (STI). However, they face challenges aligning policy with practice and securing gender-related funding (Jackson & Mackevicius, 2024). Women in STEM also encounter difficulties stemming from inflexible work systems and traditional cultural norms within Africa and abroad (Kemechian et al., 2023). To effectively monitor progress toward SDG 5 in Africa, the Afrobarometer has been proposed using UN Tier 1 indicators. However, data gaps in crucial gender inequality metrics persist across the continent, emphasizing the need for sustained efforts to address disparities and promote women's participation in STEM fields to fully achieve SDG 5 (Nhamo et al., 2018).

Gender disparity is quite prevalent in African countries, as depicted in Figure 9.3. The chart compares the female-to-male mean years of education and labor force

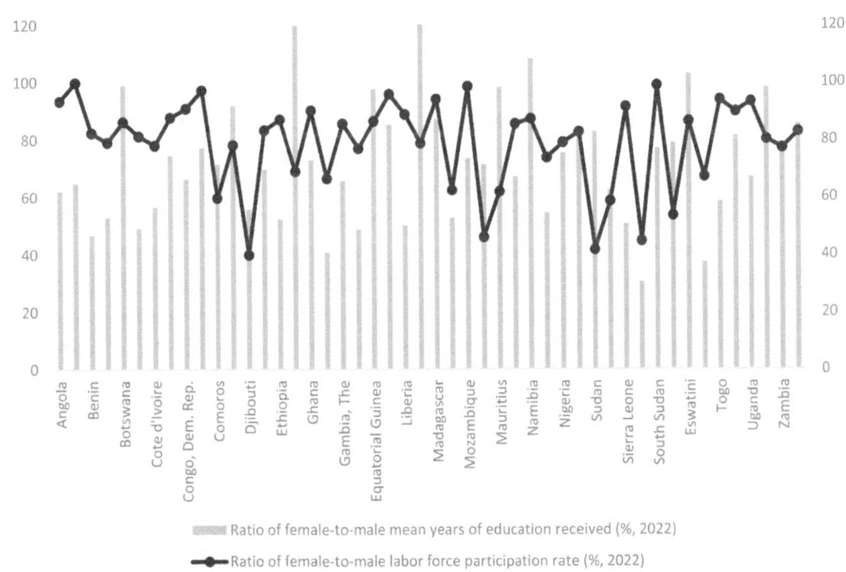

FIGURE 9.3 African female-to-male mean years of education received and labor force participation rate in 2022 Author's illustration based on UNDP and ILO data

participation ratios across various African countries in 2022. The ratio of education shows substantial variation. Countries such as Botswana, Lesotho, and Namibia exceed or approach a 100% ratio, indicating that women receive the same or more years of education compared to men. Conversely, countries like Burkina Faso, Central African Republic, and Chad report much lower ratios, with women receiving significantly fewer years of education—around 50% to 60% of their male counterparts. South Sudan and Comoros also show particularly low ratios, below 40%.

Most African countries maintain relatively consistent figures regarding the labor force participation ratio, generally between 70% and 90%. For example, Gabon, Namibia, and South Africa show participation rates close to 80%. However, a few countries, such as Eritrea and Comoros, display much lower female-to-male participation. In contrast, countries like Lesotho show higher ratios, indicating relatively more robust female participation in the workforce.

Interestingly, some countries like Mauritania and Somalia exhibit moderate labor force participation ratios despite lower female education ratios, highlighting the complex relationship between education and labor participation. This data underscores ongoing gender disparities in education and employment, which is crucial for informing gender equality policies.

BOX 2: DR AMINA ABUBAKAR: A TRAILBLAZER IN ADVANCING SDG 5 THROUGH STEM

Dr Amina Abubakar, a distinguished cognitive neuroscientist, has made significant strides toward achieving SDG 5 (Gender Equality) through her impactful work in health sciences and advocacy for women in STEM. Her research, which helps close the gender gap in scientific study, focuses on vulnerable African communities and specializes in child development and mental health. In addition to actively mentoring young women scientists, Dr Abubakar challenges prejudices that prevent women from pursuing careers in STEM domains and advocates for gender parity in academic and professional settings.

Creating psychological tests that are culturally relevant for evaluating early infant development in low- and middle-income nations is a crucial aspect of her work. She ensures accurate assessments of developmental issues faced by children, especially those afflicted by HIV, malnutrition, and malaria, by designing measures specifically for African populations. This work improves child health outcomes and empowers mothers and caregivers with the resources needed to support their children's development, linking women's empowerment with better family health.

Additionally, Dr Abubakar is committed to helping African scientists—especially women—develop their capacities. As a mentor to PhD students, she has fostered a new generation of female researchers in Kenya, South Africa, Tanzania, and Zambia. Her advice fosters women's leadership in research and public health by assisting them in overcoming the obstacles associated with STEM vocations.

Dr Abubakar advocates for policies that address gender imbalances and enhance child health at the Kenya Medical Research Institute and Pwani University. Her study emphasizes the link between the mental health of mothers and the development of their children, highlighting the necessity of gender equality in achieving comprehensive health outcomes.

Recognized with the Royal Society Pfizer Award in 2016, Dr Abubakar's contributions to psychological research inspire women in science, highlighting the critical role of female representation in advancing health and development in Africa.

Sources: Africa Connect 3; Vyas-Doorgapersad (2023)

SDG 9: Industry, Innovation, and Infrastructure

SDG 9 (Industry, Innovation, and Infrastructure) focuses on sustained, inclusive, and sustainable economic growth, full and productive employment, decent work for all, and developing quality and affordable infrastructure, especially in developing nations (Küfeoğlu, 2022). It fosters the importance of growth, innovation, and technology research to overcome problems such as climate change and inequality (Kynčlová et al., 2020). With a focus on technological advancement and infrastructure, SDG 9 aims to enhance sustainable and inclusive industrialization for continuous economic growth.

The advancements in industrialization and infrastructure development have equally benefited all the various sections of the population in improving their well-being. Infrastructural development is essential for enhancing the impact of gender equality since women can access services and compete for other opportunities. African Women in STEM services are critical in realizing SDG 9 as much as they support technological advancement, enhancement of the infrastructure, and industrialism. They bring in different ways of perceiving issues affecting decision-making in fields such as engineering, information technology, and renewable energy. By removing barriers restricting women in STEM fields, African women foster the innovation potential and decree on elaborating more permissive industrial policies. Not only do they improve gender parity, but they also advance Africa's technological standards to create efficient and sustainable growth in all sectors. Hence, there is a need for gender-sensitive infrastructure to help bridge the existing gender gap in the utilization and employment of infrastructure because women hold considerable potential to contribute to investments in green and digital solutions. However, the question is how and when these challenges exist in skills deficits, implicit bias, and discrimination.

In the contemporary global landscape, technological innovations are pivotal in driving progress, and internet connectivity is critical for achieving such advancements. As a continent, Africa has been trailing behind in terms of internet penetration. The 2022 SDG Report indicates that Seychelles leads the African region, with 86.67% of its population accessing the internet (Figure 9.4). This connectivity

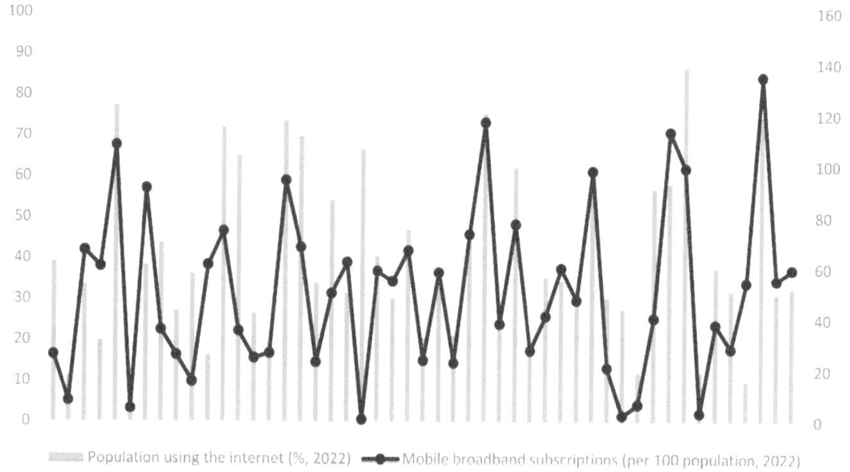

FIGURE 9.4 Percentage of African people in 2022 having mobile network and using internet in general Author's illustration based on ITU data

encompasses both fixed and mobile networks. Conversely, nations such as Burundi and Uganda report the lowest internet usage, with only 8.33% and 10% of their populations connected, respectively.

While internet access remains challenging across the continent, mobile network connectivity shows slightly more promising trends. South Africa and Mauritius are at the forefront of mobile network penetration in the region. However, several countries remain significantly underserved in this regard. For instance, Equatorial Guinea, Somalia, Chad, and the Central African Republic have some of the lowest levels of mobile network access, with only 0.9%, 2.53%, 3.4%, and 5.22% of their populations connected, respectively. These disparities highlight the urgent need for targeted interventions to expand digital infrastructure and improve internet accessibility across Africa, essential for fostering economic growth and innovation.

A nation's research and development (R&D) investment is a crucial driver of innovation and infrastructure development, as shown in Figures 9.5 and 9.6. One notable consequence of low investment in education, particularly the high dropout rates following secondary school, is the limited production of peer-reviewed, citable documents in many African countries (Figure 9.5). Seychelles leads the region with 0.86 citable documents per 1,000 people, while the majority of other countries fall significantly below 0.5. This shortfall highlights a broader lack of emphasis on academic output and knowledge creation, further reflected in the continent's generally low R&D expenditure.

Figure 9.6 illustrates the proportion of each African nation's GDP allocated to research and development. Rwanda emerges as the leader, dedicating 0.76% of its GDP to R&D, signifying a more substantial commitment to fostering innovation. In stark contrast, countries like Mauritania and Madagascar allocate a mere 0.01%

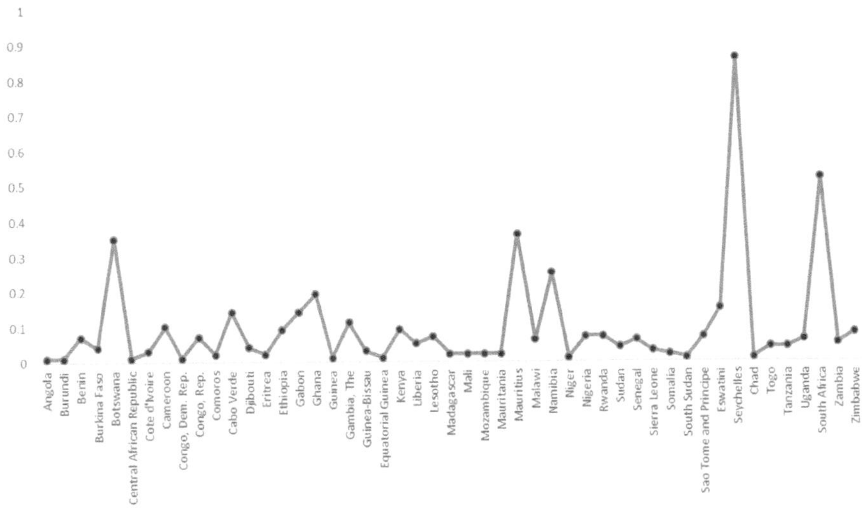

FIGURE 9.5 Number of academic articles published in journals in 2022 Author's illustration based on Scimago Journal Rank data

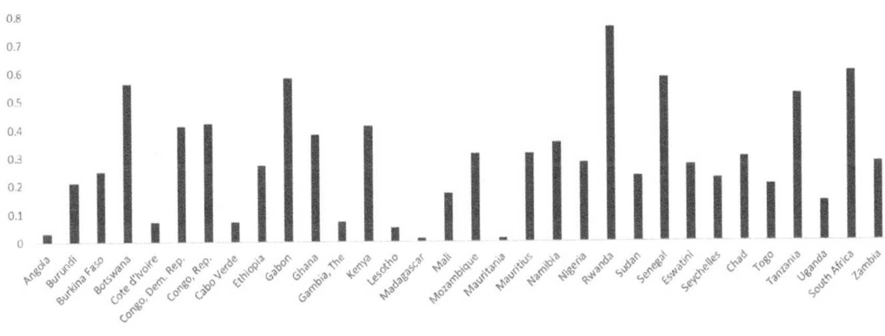

FIGURE 9.6 Expenditure on research and development (% of GDP, 2022)

of their GDP to R&D, reflecting a severe underinvestment in this critical area. This disparity underscores African countries' broader challenge in advancing scientific research and development, limiting their ability to drive technological progress and economic growth. Increasing investment in education and R&D is essential to enhance innovation and infrastructure.

The Logistic Performance Index (LPI), which assesses the quality of trade and transport infrastructure, including ports, roads, railroads, and information technology, reveals significant challenges across most African nations. On a scale ranging from one (worst) to five (best), most African countries struggle to achieve optimal performance in these critical areas, which are vital for economic growth and

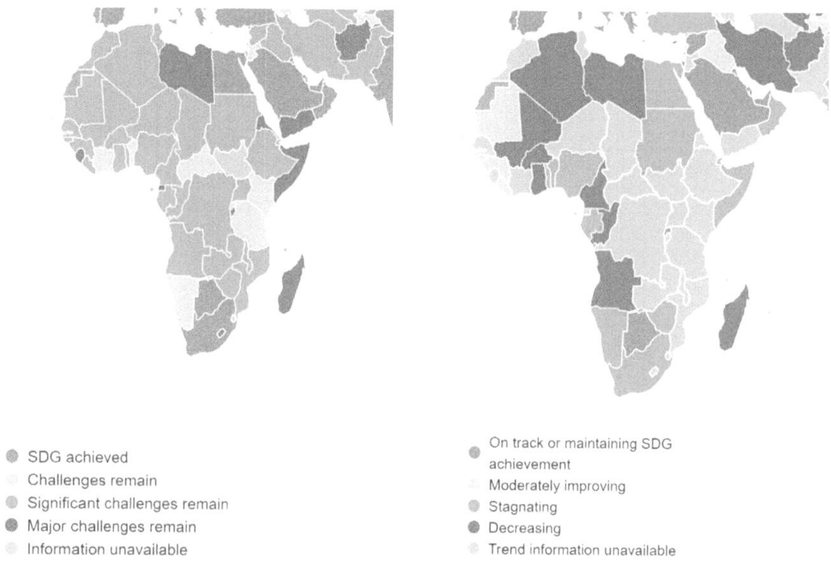

FIGURE 9.7 (a): Rating of Logistics Performance Index: Infrastructure Score in 2022 (b): Trend of Rating of Logistics Performance Index: Infrastructure Score toward 2030 Author's illustration based on World Bank data; SDG Index

international trade. According to Figure 9.7(a), only South Africa and Botswana have successfully met the SDG targets related to infrastructure quality.

However, a concerning trend emerges when examining the trajectory of the LPI scores over time. As depicted in Figure 9.7(b), many African countries are either experiencing stagnation or declining performance, indicating difficulties in improving trade and transport infrastructure. Botswana stands as an exception, demonstrating consistent progress, while the overall trend paints a bleak picture of the continent's ability to achieve the relevant SDG targets. The lack of infrastructure development in most countries not only hinders trade competitiveness but also poses a barrier to economic growth, further emphasizing the need for targeted investments and policy interventions to improve the continent's logistics performance.

BOX 3: FUNKE OPEKE—EXPANDING DIGITAL INFRASTRUCTURE

Funke Opeke has made significant strides in advancing SDG 9, emphasizing resilient infrastructure, inclusive and sustainable industrialization, and innovation. As the founder and CEO of MainOne, a leading telecommunications company in West Africa, Opeke has been instrumental in enhancing the region's digital infrastructure, particularly in Nigeria.

Some of her accomplishments include promoting West Africa's first open-access submarine cable system, including Nigeria, Ghana, and Portugal. This underwater fiber optic cable, stretching up to 7,000 kilometers in this part of Africa, was commissioned in 2010 at an overall cost of $240 million and has dramatically enhanced internet accessibility in the region. MainOne has brought high-capacity bandwidth, which has helped local technology companies and startups expand their operations and continue introducing new products into the market.

Opeke also took her efforts a notch higher by establishing MDXi, West Africa's largest Tier III data center, which provides dependable data housing and computation services. This infrastructure is essential in enhancing industrial development and technology expansion in Nigeria and other parts of the world.

Besides, there have been advancements in digital connection from the initiatives by Opeke, which have also offered employment chances in huge numbers. MainOne, through employing several people in technical positions and partnerships with other technology firms such as CcHub and Andela, has also indirectly created many employment opportunities. These partnerships have enabled startups to leverage MainOne's digital infrastructure to test and scale their business models.

Opeke's influence extends beyond corporate achievements. She chaired the Presidential Committee responsible for developing Nigeria's National Broadband Plan (2020–2025) to expand internet access nationwide. This initiative aligns with SDG 9 by promoting sustainable industrialization through enhanced connectivity.

Through her leadership at MainOne and active engagement in broadband policy, Funke Opeke has played a pivotal role in advancing SDG 9. Her efforts have improved regional connectivity and contributed to economic growth and innovation in West Africa.

Sources: Harvard Business School; Co-Creation HUB; Internet Governance Forum; Techpoint

SDG 13: Climate Action

SDG 13 (Climate Action) envisages actions to address climate change mitigation, including measures to reduce greenhouse gas emissions and adapt to its effects, as well as include climate actions within the countries' policies. SDG 13 mainly focuses on climate change mitigation through adaptation, disaster risk reduction, and sustaining development for climate-vulnerable groups. It also stresses the international partnership, funding, and capacity development in developing countries concerning climate issues. To the extent that the goal aspires to improve ecosystems, economies, and human health, there is a need to ensure that SDG 13 is achieved.

The standalone importance of African Women's STEM to SDG 13 can be seen in pursuing proper research and innovation in a climate change solution. By working in environmental science, engineering, and renewable energy sectors, they contribute to creating new technologies that address climate change and its consequences. Through advocating for sustainable and green innovation, the women in STEM advocate for breaking gender barriers while promoting climate change policies in Africa. They also make sure that women who suffer most from climate change are able to contribute to the necessary change in Africa's environmental policies. For example, the Women in Environmental Sciences (WiES) discusses how women have been instrumental in the environmental sciences and foregrounds everyone's perspectives on environmental change initiatives (Medupin, 2020). A study by Kovaleva et al. (2023) reveals increasing women's involvement in climate change decision-making. However, there are challenges, including low awareness and political willpower, that should be addressed. The authors argued that the COVID-19 pandemic has positive and negative impacts on achieving gender-balanced leadership and empowerment in climate change.

Africa is particularly vulnerable to harsh weather, food shortages, and economic instability, all of which have a disproportionately negative impact on livelihoods and the continent's ability to develop, making climate action essential. Figure 9.8 shows

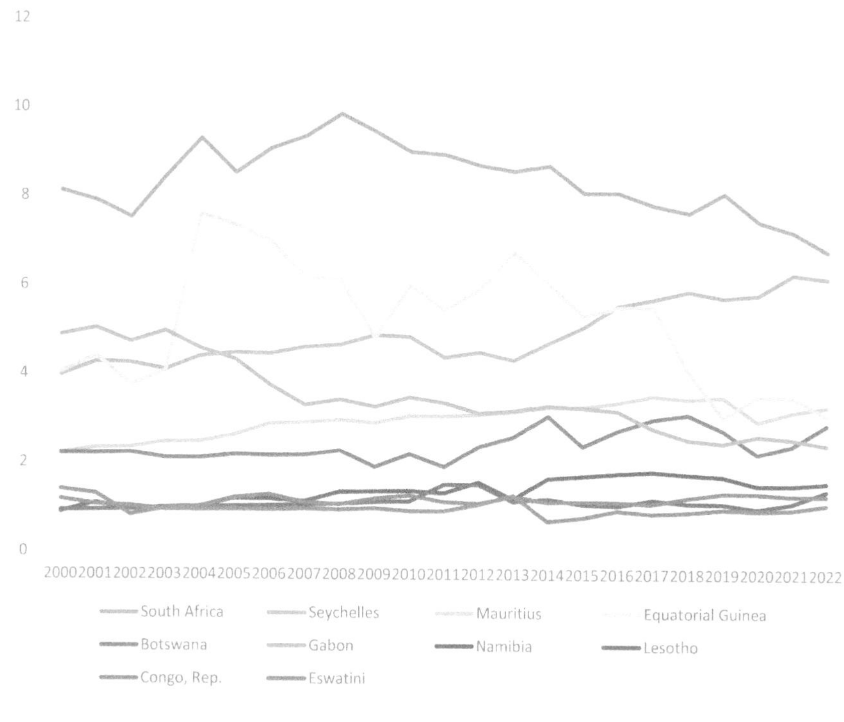

FIGURE 9.8 CO_2 emissions from fossil fuel combustion and cement production per capita from 2000 to 2022 Author's illustration based on Global Carbon Project data

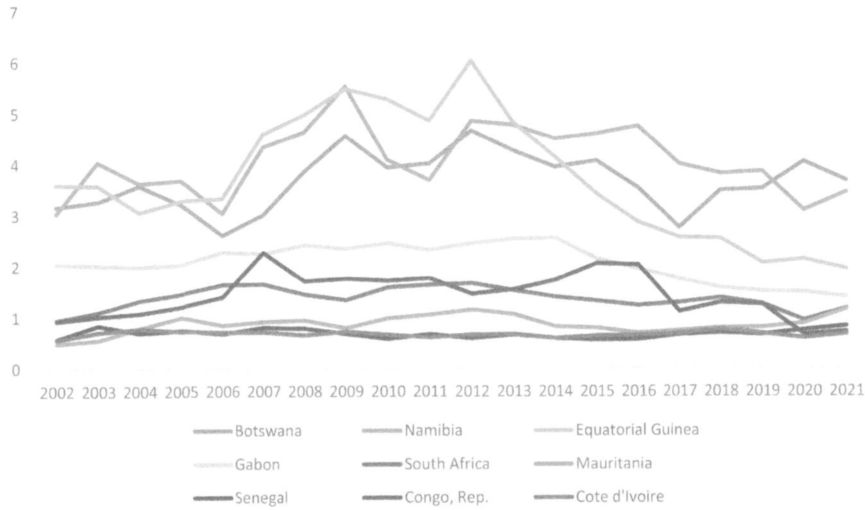

FIGURE 9.9 GHG emissions embodied in imports per capita from 2002 to 2021 Author's illustration based on Global Carbon Project data

the highest CO_2 emissions per capita in African countries between 2000 and 2022, resulting from the combustion of fossil fuels and the production of cement. South Africa consistently exhibits the highest CO_2 emissions, ranging from approximately 8 to 10 tons per capita, though there is a gradual decline post-2015. Seychelles, Equatorial Guinea, and Botswana follow, with Equatorial Guinea showing a significant spike in emissions around 2008 before declining steeply. Equatorial Guinea's emissions peaked in 2008 and then sharply decreased. Lesotho and Namibia remain at the lower end of the emission spectrum, whereas Mauritius and Gabon exhibit generally consistent trends and retain moderate emissions levels. The graph, taken as a whole, shows stark differences in emissions between African countries, with some sustaining high levels and others seeing only slight rises over time.

Figure 9.9 shows African nations' primary greenhouse gas (GHG) emitters from 2002 to 2021, as represented by imports per capita. The most significant values are continuously found in Namibia and Equatorial Guinea, peaking between 2009 and 2011 before a discernible fall. South Africa, Botswana, and Gabon display moderate but changing emission levels, with Botswana climbing post-2015. While GHG emissions from imports are lower and steadier in nations like Mauritania, Senegal, Congo, and Côte d'Ivoire, Gabon's emissions are still relatively stable. The graph shows notable differences between nations, illustrating various import trends and their environmental effects during two decades.

Additionally, most of the African countries have achieved the SDG target regarding their CO_2 emissions embodied in the exports of coal, gas, and oil, estimated by taking the five-year average of exports of fossil fuels and equating them with CO_2 emissions as depicted in Figure 9.10. Each fossil fuel's exports are limited to the

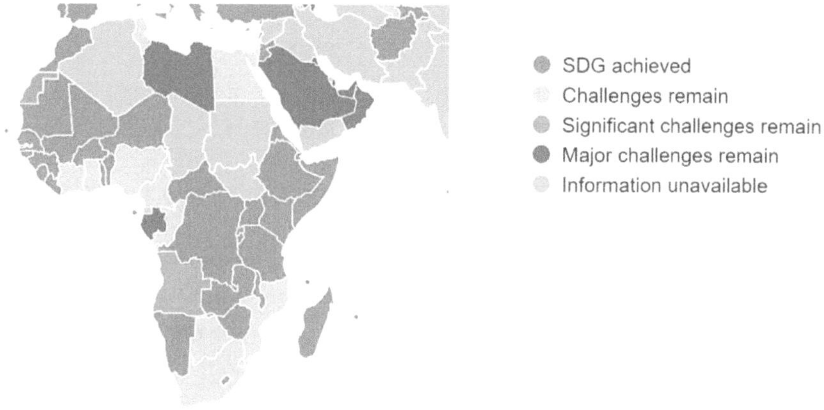

FIGURE 9.10 CO_2 emissions embodied in fossil fuel exports in 2022 Author's illustration based on UN Comtrade data; SDG Index

amount produced in the nation. However, major challenges remain in countries like Gabon and Libya, followed by Angola.

DR CATHERINE NAKALEMBE: ADVANCING CLIMATE ACTION IN AFRICA

Through her contribution and work interest in food security and agro-product resilience, Dr Catherine Nakalembe stands as a leading figure in realizing SDG 13. Currently serving as an Assistant Professor at the University of Maryland and as Africa Director for NASA Harvest, she applies satellite remote sensing and artificial intelligence to complement agriculture in Africa.

These crop monitoring techniques offer relevant information to smallholder farmers on the effects of climate change, like droughts and floods. By integrating satellite data into disaster risk financing programs, Dr Nakalembe has assisted more than 300,000 Ugandan households in obtaining food and remaining food secure in the long run. Through her work, she has impacted approaches to eliminating food insecurity worsened by climate change, hence being honored with the Africa Food Prize and the Al Sumait Prize for African Development.

Besides, Dr Nakalembe supervises capacity enhancement programs that enhance satellite data usage by local governments and organizations to develop suitable measures. Her work cuts across disciplines for multi-stakeholder partnerships, which seek to support innovation for climate-smart agriculture that addresses the region's food insecurity and climate change.

Sources: Catherine Nakalembe.com; Space in Africa; Forbes; Nakalembe et al. (2022)

The linkage between the SDGs and women in STEM is profound. By enhancing educational outcomes, promoting gender parity, driving innovation, and addressing environmental issues, women in STEM contribute significantly to the achievement of multiple SDGs. Ensuring their full and equal participation in STEM fields is essential for creating a sustainable and inclusive future.

POLICIES AND INITIATIVES SUPPORTING WOMEN IN STEM

Overview of Governmental and Non-Governmental Policies in Africa

In Africa, the government and non-governmental organizations (NGOs) have established numerous policies to support the inclusion of women in STEM. Some government initiatives include the African Union's Science, Technology, and Innovation Strategy for Africa 2024 (STISA-2024), which seek to reduce the gender gap by encouraging the enrollment of girls in STEM education. Many national policies exist; for example, in South Africa, one such initiative offers Women in Science Awards to encourage women to take up research and innovation. The African Women in Agricultural Research and Development (AWARD) program, among other NGOs, serves as a mentor and offers capacity-building opportunities to women. Despite the promising advances, gender bias and disparities in access to resources continue to present formidable barriers to integration.

Successful Initiatives and Programs

Several activities have been crucial in advancing African women in STEM fields. Women in Tech Africa, which extends to 30 countries, has been involved in initiating networking sessions and mentorship for women in the field. So far, AWARD has supported and built leadership and scientific capacity for over 1,500 African women in science and agriculture. In Kenya, there is AkiraChix, a program that trains young women in coding and other technical areas to get jobs in the technology industry. The African Network of Women in Astronomy and STEM for GIRLS in Ethiopia aim to improve girls' and women's engagement in astronomy and science (Pović, 2019). Policies encourage the proper representation of women in STEM by fostering an appropriate culture and combating impediments to engagement (Moss-Racusin et al., 2021). The African Research Academies for Women (ARA-W) summer research program works as a mentorship training program for women in STEM disciplines to erase their deficiency in Africa. As of 2014, 65 fellows have done projects in different fields of STEM, hoping to expand the program and even create an online mentorship platform (Commodore-Mensah et al., 2020). These programs have effectively removed hurdles regarding the enrollment of women in STEM fields by promoting women-friendly settings and enhancing women's engagement within the STEM fields for social and economic growth in Africa.

Recommendations for Policy Improvements and Future Strategies

In order to increase women's representation in STEM, African governments and NGOs need to develop adaptive and sustained approaches. Policies should enhance

STEM education at primary and secondary levels, inclusive of gender-sensitive curricula. Financial constraints can be eliminated by increasing scholarship programs for women in STEM disciplines. Also, the implementation of proper policies that support changes to flexible work environments, as well as anti-discrimination policies, would assist in keeping women in STEM careers. Mentorship and role model programs must be extended to support rural and marginalized areas. Last but not least, deploying public-private partnerships will integrate subject matter experts into the policymaking process so that the policies are updated regularly relative to technological advances and current trends.

CHALLENGES AND OPPORTUNITIES

Ongoing Challenges Faced by African Women in STEM

The challenges that affect STEM-related careers in Africa and its women include but are not limited to negative cultural attitudes and beliefs and limited education and resource opportunities. Cultural norms make women feel that they must be homemakers and neglect job opportunities, especially those in engineering or technology. Also, female students are not readily encouraged to take STEM education because of the limited number of female teachers and role models. In South Africa, for instance, although policies have been implemented to ensure more women enroll in STEM-related courses than ever before, African women are still constrained by race, gender, and class impediments, as per the study by Idahosa and Mkhize (2021). Regarding financing, research papers, and grants, men dominate the picture as well as there has been gender discrimination in employment and promotion. These challenges confine women not only to be fewer in STEM careers but also to be able to initiate innovations that would help transform Africa's economy.

Opportunities for Growth and Increased Impact

Despite these obstacles, African women in STEM fields have never-before-seen chances to change the world. Women inventors have more opportunities due to the IT industry's explosive growth and the growing demand for sustainable solutions to food insecurity and climate change. Particularly for women living in remote areas, initiatives like digital education and mobile learning are filling gaps in STEM education. Additionally, international organizations are giving diversity in research financing more priority, which opens doors for initiatives directed by women. Africa can spearhead technical improvements and sustainable development by using the continent's expanding youth population and supporting women's abilities in STEM fields.

Strategies to Overcome Barriers

To eliminate or mitigate these obstacles, several strategic measures are needed. Education cannot continue to operate without gender-sensitive rules. These policies should include providing mentorship and scholarships to women pursuing careers in

STEM. Therefore, altering society's culture requires increasing awareness of gender equity in the workplace and schools. Businesses can do more to support women in STEM fields by providing flexible work schedules, childcare options, and anti-discrimination measures. Governments and non-governmental organizations should collaborate to ensure that various institutions have equal access to funds for research and entrepreneurial opportunities, among other things. According to Babalola et al. (2021), accomplished African women occupying leadership roles in STEM should underscore the significance of education, diligence, and encouraging institutional frameworks for their professional growth. Aspiring African women in STEM might be inspired by sharing other women leaders' responsibilities and success stories (Babalola et al., 2021). Finally, establishing robust networks and implementing role model initiatives can motivate young women by giving them direction and chances to excel in STEM fields.

FUTURE DIRECTIONS AND RECOMMENDATIONS

Future Outlook for Women in STEM in Africa

There is a ray of hope for African women in STEM since there is increasing awareness and policy measures to address the issue. As different industries get shaped by digitalization, there will be a higher need in professions related to technology, data, and engineering that women can and should fill. Additional actions to promote women in STEM continue to be implemented with the support of governments and international organizations. However, long-term results can only be achieved where more investments are made in gender equality education, girls and women mentorship, and workplace gender equality. The ongoing trends suggest African women may assume leadership roles in innovation in renewable energy, health, and agriculture technologies.

Potential Areas for Further Research and Development

More research is still warranted in areas where African women can make an excellent impact in STEM areas. For instance, renewable energy offers women the prospect of participation in the economically lucrative and socially transformative area of sustainable energy for Africa, which is faced with dual threats of climate change and unsustainable development. The findings of gendered barriers research in STEM education could deepen our understanding of the educational models and the economic potential of women in technology entrepreneurship research; the analysis can assist in developing policy support for gender equity causes. Further, integrating STEM with health care and agricultural sectors, especially with the help of technology and software such as AI and machine learning for women, is another potential research and solution-generating sphere.

Recommendations for Stakeholders

To ensure women's engagement in STEM, governments should implement and adopt gender-sensitive policies to encourage women to engage in STEM; the policies should include scholarships, grants, and STEM-focused curricula at all levels of education. It is also necessary to promote NGOs as an essential tool to extend the possibilities of women's mentorship and support programs to help them through STEM education and business. Thus, schools as organizational systems should promote gender equality through gender-sensitive systems and gender-sensitive teaching and learning approaches. In this case, the public and private sectors should combine their efforts to ensure everyone gets a fair chance and opportunity to access the various funds and resources so that more women go for STEM-related careers. Likewise, appropriate strategies should be made to open opportunities for women in STEM, particularly for those women aspiring for leadership positions and decision-making and policy-shaping bodies in science and technology fields.

CONCLUSION

Mainstreaming African women in STEM is not solely a gender issue but is critical to attaining the SDGs. Women's perspective and creative thinking are the essentials to face today's complicated challenges like climate, healthcare services, and cities of the future. Research shows that organizations that employ more women in STEM are financially more successful and more innovative than other organizations, making gender parity the right course of action and the financially responsible one.

Therefore, the underrepresentation of African women in STEM impedes attaining the developed world's vision for the SDGs before 2030. Some of the challenges include gender stereotyping, lack of role models, and poor policy support preventing women from fully participating. These challenges must be overcome through policy change, funding female STEM education, and making workplaces more inclusive. Increasing female participation in STEM disciplines will go a long way in mending the shortage of skilled workers, positively impacting global economic growth, and providing role models to young girls and women.

Therefore, there is a need to promote women's involvement in STEM to spur more economic growth, technology, and sustainable development in Africa. To achieve the SDGs and provide women equal opportunities, the federal government, schools, and businesses should collaborate to promote women's empowerment and include women as equal stakeholders in society.

REFERENCES

Allen, C., Metternicht, G., & Wiedmann, T. (2018). Initial progress in implementing the Sustainable Development Goals (SDGs): A review of evidence from countries. *Sustainability Science*, *13*, 1453–1467.

Amunga, J. (2020). Benchmarking practices among single sex boarding secondary schools in Western Kenya. *Technium Social Sciences Journal*, *8*, 117.

Babalola, O. O., du Plessis, Y., & Babalola, S. S. (2021). Insight into the organizational culture and challenges faced by women STEM leaders in Africa. *Social Sciences*, *10*(3), 105.

Batinoluho, L. (2024). Exploring factors affecting astro-tourism development in Tanzania. *Journal of Tourism, Culinary, and Entrepreneurship (JTCE)*, *4*(1), 33–49.

Beede, D. N., Julian, T. A., Langdon, D., McKittrick, G., Khan, B., & Doms, M. E. (2011). Women in STEM: A gender gap to innovation. *Economics and Statistics Administration Issue Brief, 04–11.*

Biermann, F., Kanie, N., & Kim, R. E. (2017). Global governance by goal-setting: the novel approach of the UN Sustainable Development Goals. *Current Opinion in Environmental Sustainability, 26*, 26–31.

Blume-Kohout, M. E., & Adhikari, D. (2016). Training the scientific workforce: Does funding mechanism matter? *Research Policy, 45*(6), 1291–1303.

Commodore-Mensah, Y., Shokunbi, S., Okocha, A., Omojaro, O., Oguntuyo, S., Aneke, J., Ghanney, E. C., Huang, M., Asante, J., & Joseph, P. V. (2020). Bridging the gender equality gap in STEM to transform the pipeline in Africa: the African Research Academies for Women (ARA-W) summer research programme. *The Lancet Global Health, 8*, S40.

Crespo, B., Míguez-Álvarez, C., Arce, M. E., Cuevas, M., & Míguez, J. L. (2017). The sustainable development goals: An experience on higher education. *Sustainability, 9*(8), 1353.

Fatourou, P., Papageorgiou, Y., & Petousi, V. (2019). Women are needed in STEM: European policies and incentives. *Communications of the ACM, 62*(4), 52.

Fisher, M., Nyabaro, V., Mendum, R., & Osiru, M. (2020). Making it to the PhD: Gender and student performance in sub-Saharan Africa. *PLoS One, 15*(12), e0241915.

Fonseca, L. M., Domingues, J. P., & Dima, A. M. (2020). Mapping the sustainable development goals relationships. *Sustainability, 12*(8), 3359.

Grainger-Brown, J., & Malekpour, S. (2019). Implementing the sustainable development goals: A review of strategic tools and frameworks available to organisations. *Sustainability, 11*(5), 1381.

Henderson, K., & Loreau, M. (2023). A model of sustainable development goals: Challenges and opportunities in promoting human well-being and environmental sustainability. *Ecological Modelling, 475*, 110164.

Idahosa, G. E., & Mkhize, Z. (2021). Intersectional experiences of black South African female doctoral students in STEM: Participation, success and retention. *Agenda, 35*(2), 110–122.

Ike, M., Donovan, J. D., Topple, C., & Masli, E. K. (2019). The process of selecting and prioritising corporate sustainability issues: Insights for achieving the sustainable development goals. *Journal of Cleaner Production, 236*, 117661.

Jackson, C. K., & Mackevicius, C. L. (2024). What impacts can we expect from school spending policy? Evidence from evaluations in the United States. *American Economic Journal: Applied Economics, 16*(1), 412–446.

Kemechian, T., Sigahi, T. F. A. C., Martins, V. W. B., Rampasso, I. S., De Moraes, G. H. S. M., Serafim, M. P., Leal Filho, W., & Anholon, R. (2023). Towards the SDGs for gender equality and decent work: investigating major challenges faced by Brazilian women in STEM careers with international experience. *Discover Sustainability, 4*(1), 11.

Kiernan, L., Walsh, M., & White, E. (2023). Gender in technology, engineering and design: factors which influence low STEM subject uptake among females at third level. *International Journal of Technology and Design Education, 33*(2), 497–520.

Konyeha, S., Agwam, G. I., Musa, E., Ngonadi, I. V, & Afehomo, A. C. (2021). Initiatives and role of women scientist forums with mentorship opportunities in STEM. *African Journal of Health, Safety and Environment, 2*(2), 89–99.

Kovaleva, M., Leal Filho, W., Borgemeister, C., & Komagaeva, J. (2023). Central Asia: Exploring insights on gender considerations in climate change. *Sustainability, 15*(16), 12667.

Küfeoğlu, S. (2022). SDG-9: Industry, innovation and infrastructure. In *Emerging Technologies: Value Creation for Sustainable Development* (pp. 349–369). Springer.

Kynčlová, P., Upadhyaya, S., & Nice, T. (2020). Composite index as a measure on achieving Sustainable Development Goal 9 (SDG-9) industry-related targets: The SDG-9 index. *Applied Energy*, *265*, 114755.

Medupin, C. (2020). Women in environmental sciences (WiES) and the UN SDGs: A catalyst for achieving a sustainable future for all. *Sustainability*, *12*(17), 7116.

Mishra, M., Desul, S., Santos, C. A. G., Mishra, S. K., Kamal, A. H. M., Goswami, S., Kalumba, A. M., Biswal, R., da Silva, R. M., & Dos Santos, C. A. C. (2024). A bibliometric analysis of Sustainable Development Goals (SDGs): A review of progress, challenges, and opportunities. *Environment, Development and Sustainability*, *26*(5), 11101–11143.

Moss-Racusin, C. A., Pietri, E. S., van der Toorn, J., & Ashburn-Nardo, L. (2021). Boosting the sustainable representation of women in STEM with evidence-based policy initiatives. *Policy Insights from the Behavioral and Brain Sciences*, *8*(1), 50–58.

Mutiku, J. K., & Kiaritha, H. (2022). *Increasing the Enrolment of Women and Girls in TVET in Africa through the Women in Technical Education and Development (WITED).*

Mutsvangwa, A., & Zezekwa, N. (2021). STEM Education: A Ray of Hope for African Countries. *Unnes Science Education Journal*, *10*(2), 79–89.

Nakalembe, C., Makabe, E., & Jarvis, I. (2022). *Integrating Earth Observations into the Formulation and Implementation of National Adaptation Plans: Agriculture and Food Security GEO Supplement to the UNFCCC NAP Technical Guidelines.* https://doi.org/10.13140/RG.2.2.15991.09129

Nhamo, G., Muchuru, S., & Nhamo, S. (2018). Women's needs in new global sustainable development policy agendas. *Sustainable Development*, *26*(6), 544–552.

Nondo, T. R., & Mbereko, A. (2020). Critical analysis of the contribution of women's University in Africa towards the attainment of SDG 5. *Scaling up SDGs Implementation: Emerging Cases from State, Development and Private Sectors*, 155–167.

Ogunsanya, K. (2007). Qualifying women's leadership in Africa. *Conflict Trends*, *2007*(2), 50–54.

Okwakol, M. N., Ujeyo, M. S. S., Atibuni, D. Z., Biira, S., & Waako, P. (2022). *Overcoming policy and practice fragility and enhancing security of science, technology and innovation educational achievement for females in Uganda.*

Panigrahi, M. R., & Bhandigadi, P. (2020). *Handbook on.*

Pham, N., & Triantis, A. J. (2015). Success in STEM & the US Economy: Why the Contributions of Women Matter. *Robert H. Smith School Research Paper No. RHS*, *2877538*.

Pović, M. (2019). Development of astronomy research and education in Africa and Ethiopia. *Proceedings of the International Astronomical Union*, *15*(S367), 24–27.

Richardson, S. S., Reiches, M. W., Bruch, J., Boulicault, M., Noll, N. E., & Shattuck-Heidorn, H. (2020). Is there a gender-equality paradox in science, technology, engineering, and math (STEM)? Commentary on the study by Stoet and Geary (2018). *Psychological Science*, *31*(3), 338–341.

Sachs, J. D. (2012). From millennium development goals to sustainable development goals. *The Lancet*, *379*(9832), 2206–2211.

Schmader, T. (2023). Gender inclusion and fit in STEM. *Annual Review of Psychology*, *74*(1), 219–243.

Schmidt-Traub, G., Kroll, C., Teksoz, K., Durand-Delacre, D., & Sachs, J. D. (2017). National baselines for the Sustainable Development Goals assessed in the SDG index and dashboards. *Nature Geoscience*, *10*(8), 547–555.

Sen, G. (2019). Gender equality and women's empowerment: Feminist mobilization for the SDGs. *Global Policy*, *10*, 28–38.

Sikhosana, H., Malatji, H., & Munyoro, A. (2023). Experiences and challenges of black women enrolled in a STEM field in a South African urban university: A qualitative study. *Cogent Education*, *10*(2), 2273646.

Stafford-Smith, M., Griggs, D., Gaffney, O., Ullah, F., Reyers, B., Kanie, N., Stigson, B., Shrivastava, P., Leach, M., & O'Connell, D. (2017). Integration: the key to implementing the sustainable development goals. *Sustainability Science*, *12*, 911–919.

Stewart-Williams, S., & Halsey, L. G. (2021). Men, women and STEM: Why the differences and what should be done? *European Journal of Personality*, *35*(1), 3–39.

Stewart, F. (2015). The sustainable development goals: A comment. *Journal of Global Ethics*, *11*(3), 288–293.

Sudarkasa, N. (1986). " The Status of Women" in Indigenous African Societies. *Feminist Studies*, *12*(1), 91–103.

Vyas-Doorgapersad, S. (2023). Challenges to achieve the goal 5 of the sustainable development goals (SDG) in South Africa. *International Journal of Research in Business and Social Science (2147-4478)*, *12*(6), 257–266.

10 The Evolving Role of Women in STEM in Africa

Cultural, Societal, and Regional Perspectives, 2024

Dativah Bideri Mukamusonera

INTRODUCTION

This chapter aims "to effulgence policymakers, financier, researchers and academia and entrepreneurs to on empowering girls and women in STEM from rural areas. It seeks to provide a road map for collectively breaking down the barriers preventing women from pursuing and succeeding in these fields, thereby contributing to a more inclusive and innovative global workforce.

The Gender Digital Divide in rural Africa is a multifaceted issue shaped by socioeconomic factors, cultural norms, and infrastructural deficits. Closing this divide requires targeted policies and initiatives that address both demand and supply-side barriers, focusing on affordability, education, and gender norms. Addressing this divide empowers girls and women and boosts overall economic growth in rural.

The persistent issue reflects the status quo on culture and biology in rural. This divide profoundly affects girls economically, education, and health limiting opportunities and rights.

Access to Digital Technologies

Infrastructure Deficit: Many rural African areas lack basic internet infrastructure such as mobile networks and broadband access. This lack of infrastructure disproportionately impacts women, who are often more likely to live in these underserved rural communities.

Gender Gap in Mobile Ownership: According to the GSM Association (GSMA), "women in Sub-Saharan Africa are 13% less likely to own a mobile phone." This translates to around 74 million fewer women owning mobile phones in the region.

Internet Usage Gap: The internet gender gap is even more significant. "Women in Sub-Saharan Africa are 41% less likely to use the internet than men," according to GSMA's "Mobile Gender Gap Report 2023." This figure highlights the extent to which women are excluded from the digital world.

DOI: 10.1201/9781003540618-10

SOCIOECONOMIC BARRIERS

Cost of Mobile and Data Services: According to a 2021 study by the Alliance for Affordable Internet (A4AI), "the cost of 1GB of mobile data represents 6.5% of monthly income for rural women, compared to 3.4% for men in the same communities." Cost is a major barrier to digital inclusion for women in rural Africa. On average, women have lower incomes than men and are less likely to afford mobile phones and data services.

Education and Digital Literacy: Low levels of digital literacy and educational attainment among women compound the divide. UNESCO estimates that "more than 50% of women in Sub-Saharan Africa are illiterate," which makes navigating digital platforms more challenging. Without digital skills, "women are less likely to use online services," further widening the gap.

CULTURAL AND SOCIAL NORMS

Gender Norms and Roles

Cultural and social norms often dictate that women's roles are centered around domestic responsibilities, limiting their freedom and time to engage with technology. In rural settings, where patriarchal structures are strong, women may face resistance from male family members regarding phone ownership or internet use.

Safety and Privacy Concerns

Women also cite concerns about safety and privacy when using digital technologies. According to the Web Foundation, women in rural areas are 50% more likely than men to face online harassment, which can deter their digital engagement.

IMPACT OF THE DIGITAL GENDER DIVIDE

Economic Disempowerment

Women's limited access to digital tools hinders their participation in economic activities. A report by the World Bank found that enhancing women's access to digital technologies could contribute an additional $12 billion to GDP annually in Sub-Saharan Africa.

Women are limited to critical information.

Limitation to digital opportunities means deprivation of critical information for modern agriculture, health, and finance. For instance, women farmers are less likely to use digital agricultural platforms, which offer information on weather patterns, crop prices, and best practices.

EFFORTS TO BRIDGE THE DIVIDE

Government and NGO Initiatives

There are multiple ongoing efforts to close the digital gender gap. For example, the Equal Access Initiative by the GSMA works with governments and mobile network operators to promote digital inclusion for women. Similarly, organizations like Digital Opportunity Trust focus on training rural women in digital skills to empower them economically.

Gender Disparities in Education, Especially in STEM Fields

While there has been progress in increasing female participation in education over the past few decades, significant gaps exist, particularly in STEM subjects. These disparities are influenced by a complex mix of cultural, societal, and regional factors, which work together to support or hinder gender equity in education.

Cultural Factors

In many cultures, traditional gender roles dictate that men should pursue careers in engineering, mathematics, and technology, while women are encouraged to choose fields deemed more suitable for their perceived "nurturing" nature, such as education or healthcare. This cultural bias can limit young girls' exposure to, and interest in, STEM subjects from an early age. Parents, teachers, and society often subtly or overtly reinforce these stereotypes, discouraging girls from pursuing paths in science and technology.

Societal Expectations and Biases

Gender biases in society perpetuate the stereotype that boys are inherently better at math and science than girls. These false expectations impact girls' confidence and interest in STEM subjects. "The under-representation of women in leadership positions in STEM fields and the lack of visible female role models in STEM careers further creates a cycle of exclusion" (UNESCO,2017).

Regional Disparities

In some regions, particularly in developing countries, girls face additional barriers to education due to economic challenges, limited access to schooling, and cultural practices such as early marriage. In such contexts, even when girls can attend school, they may face limited opportunities to engage in STEM education due to poorly resourced schools, a lack of trained teachers in science and technology, and societal pressures that prioritize boys' education over girls'.

Workplace Culture and Representation

Even when women overcome these barriers and pursue higher education in STEM fields, they often encounter gender-based challenges in the workplace. STEM-related industries can have environments that are unwelcoming or even hostile to women, who may face issues such as gender discrimination, unequal pay, lack of mentorship opportunities, and limited career advancement. These challenges lead to a significant drop in women staying and thriving in STEM careers, further exacerbating the gap.

PURPOSE OF THE CHAPTER

Highlighting the Mission of Girls Tech Ventures (GTV)

The mission of Girls Tech Ventures (GTV) is to address these gender disparities in education and career pathways, with a particular focus on STEM fields. GTV aims to empower young girls and women by providing access to STEM education, mentorship, and leadership opportunities, thereby breaking down the cultural, societal, and regional barriers that contribute to these disparities.

The organization recognizes that simply providing educational opportunities is not enough. To make a lasting impact, GTV works on multiple levels to address the underlying causes of gender inequality in STEM:

Model: Girls Tech Ventures offers a solution model that empowers rural youth centers in Rwanda, focusing on equipping women and girls with 360-degree digital skills. The initiative creates digital entrepreneurial opportunities, offering six-month short courses for teen mothers in online services; online banking, telecommunications service, money transfers, and technology transfer in agriculture to strengthen sustainable STEM.

Early Exposure and Education: GTV promotes early exposure to STEM fields for girls through after-school programs, workshops, and summer camps. Aiming to link girls with hands-on learning experiences in coding, robotics, mathematics, and engineering, GTV helps spark their interest and build confidence in their abilities from a young age.

Mentorship and Role Models: Recognizing the importance of mentorship, GTV connects girls with women leaders in STEM fields. These mentors serve as role models, offering guidance and sharing their experiences to inspire the next generation of female STEM professionals. GTV also creates a network of women in STEM, fostering community and support systems to help girls navigate challenges.

Challenging Stereotypes and Biases

GTV actively works to combat the societal and cultural biases that deter girls from pursuing STEM education. This involves outreach to parents, teachers, and communities against gender biases and encourages girls to engage in STEM subjects. GTV also advocates for policy changes in the workplace.

Regional Focus

GTV operates with a global perspective, adapting its programs to meet the specific needs of different regions. In developing countries, GTV collaborates with local organizations to improve access to education for girls and address cultural practices that hinder their participation in STEM. In more developed regions, GTV focuses on building career pathways and creating inclusive workplace cultures where women in STEM can thrive.

Challenges Faced by Rural Girls in STEM Education

Gender digital divide barriers create a complex environment that prevents many rural girls from accessing quality STEM education in Rwanda and other parts of Africa. Rural girls in Africa, including Rwanda, still face numerous challenges in accessing STEM (Science, Technology, Engineering, and Mathematics) education. Key barriers include high school dropout rates, early pregnancies, poverty, and deeply ingrained social norms.

School Dropout Rates

In Rwanda, while primary school enrollment is high, dropout rates have increased, particularly in rural areas. For instance, between 2019 and 2021, primary school dropout rates increased from 7.8% to 9.5%, and from 8.2% to 9.2% in secondary schools. The transition from primary to secondary education remains low, especially for girls, due to economic pressures and traditional gender roles. Girls are more likely to drop out during the transition to secondary school, with 20.8% leaving at the end of primary school in 2016 (UNICEF) (Hempel Foundation).

Teen Pregnancies

Teenage pregnancies are a significant barrier to education for rural girls. In Rwanda, especially during the COVID-19 pandemic, over 1,500 girls became pregnant due to increased exposure to gender-based violence. In the Eastern Province alone, over 9,000 teenage pregnancies were reported, severely impacting their ability to return to school. Once girls become pregnant, they face societal stigma, leading to early marriages or an end to their education (Hempel Foundation) (*Rwanda Today*).

Poverty and Socioeconomic Barriers

Poverty is a critical challenge, particularly in rural areas, where many families prioritize boys' education over girls'. Families may struggle to afford uniforms, fees, or school materials, causing many girls to engage in child labor to support their households. This economic burden worsens their chances of continuing education (UNICEF) (Rwanda Today).

Cultural and Social Norms

In many rural communities, traditional gender roles and societal expectations to take on domestic duties or marry early pull girls away from school or reduce the time for revisions after class hours. Efforts to address these norms, such as awareness campaigns and remedial programs targeting girls, are being implemented in some regions, but change remains slow (Hempel Foundation).

Cultural and Societal Barriers

Cultural and societal barriers to girls' access to STEM education in rural Africa reveal a range of cultural and societal barriers that disproportionately hinder girls' access to STEM (Science, Technology, Engineering, and Mathematics) fields. These barriers include traditional gender roles, early marriages, and societal expectations. Each of these factors significantly limits girls' educational opportunities, particularly in technical subjects, perpetuating the gender gap in STEM.

Traditional Gender Roles

Cultural Expectations: In many rural African societies, traditional gender roles define the primary responsibilities of women as domestic caregivers. This entrenched view limits the time and opportunity for girls to pursue education, especially in subjects like STEM, which may require long hours of study and practice.

Household Chores

Girls are often expected to contribute, reducing the time they can devote to schoolwork. Studies show that girls in rural African areas spend two to three times more hours on domestic work than boys.

Perception of STEM

STEM fields are often seen as masculine. Boys are encouraged to pursue science and technology subjects, while girls are steered toward more "feminine" subjects like home economics or arts. This gender bias creates a significant barrier to girls' interest in STEM from an early age.

Early Marriages

Prevalence of Early Marriages: In many parts of rural Africa, early marriages are common, often seen as a way to alleviate economic burdens or adhere to social norms. Reports indicate that four in ten girls are married before the age of 18. Once married, girls are expected to prioritize family responsibilities, leaving little or no time for continuing education.

Impact on Education: Early marriage usually results in girls dropping out of school. Pregnancies and the responsibilities of marriage mean returning to formal education is difficult, let alone engaging in technical or advanced studies such as STEM.

Missed Opportunities

Early marriage cuts short educational attainment for girls, leaving them with fewer skills and qualifications, which exclude them from more lucrative, skilled labor markets often associated with STEM careers.

Societal Expectations and Norms

Female Subservience: In many rural African communities, there is a cultural norm that women should be subservient to men. This societal expectation discourages girls from pursuing leadership roles, speaking out, or challenging existing gender biases, including those related to education. As a result, girls may lack the confidence to excel in fields like STEM.

Parental Attitudes: Parental attitudes often influence educational outcomes. Many parents in rural areas view investment in boys' education as a better long-term financial decision, while girls' education is sometimes seen as less important since they are expected to marry and leave the household. In a 2018 UNICEF report, it was found that parents in many African rural communities are more likely to invest in boys' education over girls, particularly in higher-cost, specialized education like STEM.

Impact of These Barriers on Girls' Access to STEM Education

Low Enrollment in STEM: Across Africa, girls make up less than 30% of students enrolled in STEM subjects. In rural areas, this number is significantly lower due to societal and cultural barriers.

Low Confidence in STEM Abilities: Due to traditional gender norms, girls are often socialized to believe they are less capable in subjects like math and science. This self-doubt, combined with external pressures, limits their engagement with STEM education.

Early Dropouts: Early marriages and societal pressure result in high dropout rates for girls in rural areas. This is particularly true in secondary school, where STEM subjects become more prominent. As a result, fewer girls graduate with the qualifications needed to pursue further education or careers in STEM fields.

Long-term Economic Implications: By limiting access to STEM education, these barriers perpetuate economic inequality. STEM careers are among the highest-paying, and the exclusion of girls from these fields contributes to the persistent gender wage gap.

Addressing the Barriers

Community Education and Awareness Campaigns: Raising awareness about the value of girls' education, particularly in STEM, can help challenge traditional gender roles. Programs like UNESCO's "STEM and Gender Advancement (SAGA)" aim to shift perceptions of STEM as a male-dominated field.

Role Model Programs: Introducing female STEM professionals as role models can inspire young girls to pursue science and technology fields. Programs that showcase successful women from rural areas can help shift the narrative.

Policy Interventions: Governments and NGOs can create policies that delay early marriages and encourage continued education for girls. Conditional cash transfers for families that keep girls in school or programs that provide scholarships for STEM education can help alleviate some financial barriers.

Mentorship and Support Networks

Providing girls with mentorship and peer support networks helps build their confidence in STEM subjects. These networks can serve as a buffer against societal pressures to conform to traditional roles.

Addressing the cultural and societal barriers to girls' access to STEM education in rural Africa requires a multifaceted approach that tackles traditional gender roles, early marriage practices, and societal expectations. By fostering a supportive environment for girls in STEM, it becomes possible to break the cycle of inequality and empower the next generation of female scientists, engineers, and innovators.

Educational and Technological Barriers

Technology

Digital Divide

Africa has a significant lack of access to digital technology, including computers, internet connectivity, and other necessary infrastructure. This is especially problematic for STEM education, which increasingly requires digital tools for learning and innovation.

Gender Disparities

Girls and women face more challenges in accessing technology due to socio-cultural norms, which may prioritize boys' education or view technology-related fields as inappropriate for women. This creates a gender gap in digital literacy and STEM engagement.

Inadequate Educational Infrastructure

Facilities and Resources: Many rural schools lack basic infrastructure such as electricity, proper classrooms, and laboratories that are crucial for teaching STEM subjects effectively. The absence of science labs, for instance, prevents practical learning experiences that are essential for subjects like physics, chemistry, and biology.

Teacher Shortage and Training: Teachers in rural schools often lack the necessary training in STEM subjects and technology. This is compounded by a general shortage of qualified teachers, making it difficult to provide high-quality education in these fields.

Limited Resources for STEM Programs

Financial Constraints: Schools in rural Africa often have very limited budgets, which restricts their ability to invest in STEM programs, buy essential teaching materials, or offer specialized training for teachers.

Curriculum Challenges: National curriculums may not prioritize STEM education in a way that fosters interest and engagement. In many cases, STEM subjects are underdeveloped, with little emphasis on practical skills, innovation, or technology use.

Gender and STEM

Focus on Women and Girls

Cultural Barriers: Cultural stereotypes and gender biases are prominent in many African societies, where girls are often discouraged from pursuing education, especially in male-dominated fields like STEM.

Early Dropout Rates: For girls in rural Africa, factors influencing dropout include early marriage, domestic responsibilities, and a lack of perceived value in continuing education.

Path Forward

Addressing these barriers requires a multifaceted approach, including:

Investment in infrastructure: Governments and international organizations should focus on building and upgrading educational infrastructure, particularly for digital tools and STEM labs.

Promoting gender equality

There should be specific policies and initiatives aimed at encouraging women and girls to pursue STEM, including scholarships, mentorship programs, and community sensitization.

Teacher training and curriculum development: A modernized, practical-oriented STEM curriculum is essential. These combined efforts can bridge the gap and ensure equitable opportunities, helping to foster a new generation of STEM professionals.

THE ROLE OF GIRLS TECH VENTURES: A GROUNDBREAKING INITIATIVE

Genesis and Vision

The founding story of Girls Tech Ventures (GTV) is deeply intertwined with the personal journey of its founder. The founding story of Girls Tech Ventures and Dativah Mukamusonera's personal journey share a powerful narrative of resilience, empowerment, and a deep commitment to overcoming gender-based inequalities.

Dativah Mukamusonera, growing up in a rural setting, faced significant challenges similar to those described in the founding story of Girls Tech Ventures. Both narratives highlight how cultural norms prioritized boys' education over girls'. In Dativah's case, her brothers were encouraged to pursue schooling while she was expected to fulfill domestic roles, eventually being forced into marriage and denied an education. The founder of Girls Tech Ventures experienced a similar fate when she was forced to leave school at a young age after the death of her father, leaving her dreams shattered.

Dativah fought for her education, eventually completing her BBA and MBA, and went on to hold leadership positions on several boards, including Gahini Hospital, NGOs, and partnering with international women's organizations like the GOA under the Obama Foundation. Similarly, the founder of Girls Tech Ventures, through sheer determination, reclaimed her right to an education by claiming her rights from the local government, an act that became a transformative moment in her life.

Dativah's work with multiple women's organizations mirrors the mission of Girls Tech Ventures to uplift girls in rural areas. Dativah recognizes that many girls continue to face the same struggles they once did and that this situation needs to change. Dativah acknowledges the critical role that technology plays in breaking the cycle of gender inequality.

Technology is a key focus for Girls Tech Ventures. Forced to navigate a male-dominated world, Dativah, who led initiatives to found Girls Tech Ventures, emphasizes that her lack of access to tech growing up in rural Rwanda motivated her to ensure that other girls don't miss out on these opportunities. Dativah understands that equipping girls with tech skills is essential to preparing them for leadership and success in the modern world, particularly as the digital divide continues to deepen for marginalized communities.

Dativah Mukamusonera's personal journey and the founding story of Girls Tech Ventures are deeply intertwined, overcoming gender-based discrimination, reclaiming the right to education, and fostering opportunities for young girls to break barriers through STEM awareness among girls in rural Africa.

Dativah says:

> Girls Tech Ventures was founded with a clear mission: to create opportunities for girls in rural Africa-Rwanda to access education, particularly in technology, STEM, and entrepreneurship. In addition to my personal story, there's another reason I focused on technology. I never had the chance to develop tech skills myself. Growing up in a rural area, access to technology and digital tools was limited, and it wasn't even seen as important for girls. But I knew, even back then, that technology could open doors to a future that wasn't available to me. This lack of opportunity for girls in tech is still a reality in many parts of rural Rwanda, and I wanted to change that.

The importance of technology became even clearer during the COVID-19 pandemic. When schools shut down and much of the world moved online, many girls in rural areas were left behind. They didn't have access to the internet, devices, or the digital skills needed to continue learning. The pandemic exposed the deep digital divide and it highlighted the urgent need for girls to have access to technology. Without it, they are at risk of falling further behind in education and missing out on opportunities that could change their lives.

This is why Girls Tech Ventures focuses so heavily on technology. It's not just about giving girls the skills to navigate the digital world—it's about giving them the tools to succeed in life. By equipping girls with technology and STEM skills, we are preparing them for a future where these abilities will be essential in nearly every field. We are empowering them to innovate and create solutions that can transform their communities.

For me, Girls Tech Ventures is more than just an organization—it is a movement to redefine what is possible for girls in rural Rwanda. My story may have started with personal struggle, but it has grown into a story of empowerment, transformation, and hope for every girl who dares to dream beyond the limitations society imposes on her. Together, we are building a future where every girl has the chance to rise, thrive, and become a leader in her own right, and where no girl is left behind due to a lack of technology or opportunity"

Holistic Approach

The Added Value of Leveraging Technology for Rural Women's Opportunities in Africa: A Holistic Approach by GTV

Women in rural Africa have long faced significant challenges in accessing economic opportunities, education, and essential services. These limitations have contributed to cycles of poverty, with restricted pathways for empowerment and autonomy. However, leveraging technology is a game-changer, providing rural women with new opportunities for economic participation, skill development, and access to services. Within the framework of GTV's holistic approach, technology uplifts rural women, transforming them from marginalized groups into empowered economic contributors. Here are five key areas where GTV creates a positive impact by integrating technology into its programs.

Empowerment Through Digital Entrepreneurship

Technology in entrepreneurship presents a vast range of opportunities for rural women. GTV's commitment to bridging the gap between rural isolation and global markets realizes digital platforms that facilitate business expansion, financial independence, and access to a larger customer base.

Access to Digital Tools for Business Expansion

Technology provides rural women with access to tools like e-commerce platforms, digital marketing, and mobile payment systems. Through digital entrepreneurship, women can transcend the limitations of local economies, selling products to customers beyond their immediate communities. For example, rural women producing handcrafted goods, textiles, or agricultural products can leverage platforms like social media, Etsy, or Amazon to reach both national and international markets. GTV facilitates training on these platforms, teaching rural women how to establish an online presence, manage logistics, and handle digital payments securely.

Training and Education on Digital Marketing

GTV offers programs that empower women to promote their businesses more effectively, reaching a wider audience. Learning how to harness the power of digital marketing helps rural women compete with larger businesses, transforming them into savvy entrepreneurs with the ability to scale their ventures.

Financial Tools and Autonomy

Access to online financial tools also promotes entrepreneurship by enabling rural women to manage their businesses more efficiently. Mobile banking platforms and digital wallets, such as M-Pesa in Kenya, provide access to savings, credit, and payment solutions without the need for traditional banks. With these tools, women can save money, invest in new business opportunities, and participate in microloan programs, further increasing their financial autonomy.

Technology-Driven Agriculture

Agriculture is a cornerstone of rural economies in Africa, and women form the backbone of the sector. However, they often face challenges related to low productivity, lack of access to resources, and the physical burden of farming. GTV's tech transfer programs, such as precision farming and climate-smart agriculture, offer innovative solutions that enhance productivity while reducing the physical demands of farming.

Decisions

Precision allows women to use technology such as GPS, drones, and sensors to inform planting, watering, or harvest crops, optimizing yields and reducing waste. These technologies can significantly improve agricultural productivity, allowing rural women to grow more food with fewer resources. GTV could partner with technology providers to deliver these tools to women farmers and offer training on how to interpret the data and make informed decisions.

Climate change is a pressing global challenge, disproportionately affecting vulnerable communities, especially in rural Africa. Women, who often manage both agricultural and domestic responsibilities, face heightened risks due to their limited access to resources for adaptation. However, integrating technology and innovative agricultural practices can significantly mitigate these impacts, transforming livelihoods and promoting long-term sustainability. Organizations like Girls Tech Ventures (GTV) have been instrumental in advancing such initiatives by providing access to education, technology, and skills development that empower women in sectors traditionally dominated by men. This article explores how GTV and similar organizations leverage climate-smart agriculture, automation, digital platforms, and mentorship to support women in rural Africa, fostering economic independence and community development.

Climate-Smart Agriculture and Gender Equity

Climate-smart agriculture (CSA) is crucial for adapting to the challenges posed by climate change, especially for small-scale rural farmers. Techniques such as the use of drought-resistant crops, water-saving irrigation systems, and agroforestry help women farmers manage variability.

Rural women have limited access to land, credit, and agricultural inputs. These constraints make it difficult for them to adopt new technologies or practices. However, targeted training and support, such as that provided by GTV, can empower women with the knowledge and tools to implement CSA practices. Through workshops and community outreach, GTV introduces these techniques, enabling women to sustain their agricultural productivity despite climate challenges. This not only secures food sources for their families but also promotes long-term environmental sustainability.

Aligning with Similar Initiatives

Similar initiatives have been launched that advocate for gender-responsive CSA. They emphasize the importance of involving women in decision-making processes related to climate change adaptation and promoting access to technologies that are tailored to their needs. Studies have shown that when women are equipped with climate-smart practices, agricultural productivity improves, benefiting both their households and the wider community (World Bank, 2015).

Automation in Agriculture

Reducing Physical Labor

The integration of automation technologies in agriculture can significantly reduce the physical labor burden on women, allowing them to manage their farms more efficiently while balancing domestic duties. Small-scale farming equipment and mechanized tools, such as handheld seeders or solar-powered irrigation systems, can lessen the physical strain of traditional farming methods. This is particularly important for rural women, who often juggle agricultural work with caregiver and household responsibilities.

By incorporating labor-saving technologies into their programs, GTV helps women not only enhance their farming efficiency but also free up time for other income-generating activities or educational pursuits. These tools enable women to scale their agricultural operations, increasing productivity and allowing them to diversify their sources of income.

Case Studies of Successful Implementation

In countries like Kenya and Ethiopia, projects introducing small-scale mechanization have shown promising results in reducing labor intensity and improving yields. A study by Doss et al. (2011) highlighted how women's labor burden decreased by 30% in communities where mechanized tools were introduced, allowing more time for entrepreneurship or education.

SCALABILITY THROUGH TECHNOLOGY PLATFORMS

Education and Skill Development

Access to education is a powerful driver of economic growth and social change, yet rural women in Africa often face geographic isolation limiting their opportunities.

Digital platforms provide a scalable solution by offering online vocational opportunities.

STEM Education for Rural Women

STEM education opens doors for women to enter high-demand industries such as technology, engineering, and healthcare. By leveraging technology, GTV provides rural women with access to online learning platforms, enabling them to acquire skills in coding, web development, and data analysis. These platforms allow women to study at their own pace and overcome barriers related to distance, time, or cost.

Vocational training programs also play a key role in equipping women with practical skills in areas like solar panel installation, smartphone repair, or digital health services. This prepares them for careers that can lead to economic independence and contribute to rural development.

Educational Models Aligned with GTV

Similar models of saleable education have been implemented by organizations, expanding digital literacy and vocational training for women in underserved regions. Research indicates that providing women with STEM education not only enhances their employability but also drives innovation and problem-solving in their communities (UNESCO, 2017).

Digital Inclusion: Enhancing Access to Resources

Water, healthcare, and sanitation remain challenges in many rural areas. Digital inclusion can bridge these gaps by providing women with tools to manage these resources more effectively. For instance, smart sensors can be used to monitor water quality and availability, reducing the time women spend fetching water and enabling better management of this vital resource.

Mobile Health and WASH Platforms

Mobile health applications (mHealth) offer women access to crucial healthcare services, such as maternal health, family planning, and nutrition guidance. These platforms connect rural women with healthcare providers and telemedicine services, improving their health outcomes and allowing them to focus on their businesses and families. Additionally, digital tools can be used to educate women on water, sanitation, and hygiene (WASH) practices, helping to reduce health risks and improve community well-being.

Example from Kenya

The M-Pesa platform in Kenya, initially designed for mobile banking, has expanded to include health and water services through partnerships with organizations like Safaricom. Women in rural areas now access healthcare and manage water resources via mobile technology, improving overall community health (Jack & Suri, 2011).

Financial Inclusion through Digital Services

Financial inclusion is critical for empowering rural women, allowing them to save, invest, and build sustainable businesses. However, traditional banking services are often inaccessible in remote areas. Digital financial platforms, such as mobile banking and microloan services, can provide women with the financial tools they need to manage their resources effectively.

Mobile Banking and Microloans

GTV promotes the use of online platforms among rural women, helping them save money, apply for microloans, and receive payments for their goods and services. Access to credit allows women to invest in their businesses, leading to increased economic stability and independence.

Digital Literacy and Financial Management

GTV also offers financial literacy training, teaching women how to manage their finances using mobile platforms. This training equips women with the skills to track expenses, plan for the future, and make informed financial decisions. Studies show that financial literacy improves not only individual outcomes but also the overall financial health of rural communities (Klapper et al., 2015).

The Multiplier Effect and Long-Term Impact

The initiatives undertaken by GTV create a multiplier effect, where advancements in one area fuel growth in others. For example, women who gain access to education and financial services can reinvest their earnings in their farms or businesses, further improving their productivity and economic stability. As women become more financially independent, they can invest in their children's education, contributing to long-term community development.

Strategic Partnerships

The success of GTV's programs is bolstered by strategic partnerships with tech companies, local governments, and NGOs. Collaborations with companies like Google and Microsoft provide essential resources, such as software tools and scholarships, while partnerships with local organizations ensure that programs are accessible even in remote areas.

Girls Coding Academy

One of GTV's most innovative programs, "Coding for Change," teaches girls to develop apps and websites that address local challenges such as waste management and water conservation. Empowering girls to become problem-solvers in their communities further amplifies the impact of GTV's initiatives.

Mentorship and Role Models

GTV's mentorship programs connect participants with successful professionals in STEM and Agri-tech, providing guidance and support as they navigate these male-dominated fields. Female mentors who have overcome similar challenges offer both technical skills and personal development advice, helping girls build confidence and resilience.

SRHR and STEM Mentorship

Incorporating mentorship in Sexual and Reproductive Health and Rights (SRHR) into GTV's programs ensures that girls have access to accurate health information and resources. This holistic approach links personal well-being with professional success, reinforcing the idea that empowered, healthy women are better positioned to succeed in their careers and advocate for their rights.

CONCLUSION

By integrating climate-smart agriculture, automation, digital education, and financial inclusion into its programs, Girls Tech Ventures is transforming the lives of rural women in Africa through strategic partnerships, scalable education platforms, and mentorship programs. These initiatives not only improve individual outcomes but also create a ripple effect that benefits entire communities.

REFERENCES

Economic Forum. *Why STEM Education for Women and Girls Is Crucial.* 2024. Why-stem-education-for-women-is-important.

Girls Tech Ventures. *Empowering Women in Agriculture through Technology and Innovation.* Accessed October 12, 2024. https://www.gtv.org/empowering-women.

Ouma, Josephine, and Anna Katharina Hornidge. *Women, Digital Technologies, and Agricultural Productivity in Rural Africa: Uganda* 2020.

Safaricom. M-PESA: *A Global Success Story.* Accessed October 13, 2024.

Yeboah, Kenneth A., and Marian Dossou-Yovo. *Climate-Smart Agriculture and Women's Empowerment* 2020.

11 Pioneering Economic Advancement through Mentorship Abstract

Rolayo Akhigbe

PIONEERING ECONOMIC ADVANCEMENT THROUGH MENTORSHIP CHAPTER

There have been many reports shared about the disproportionate amount of funding that goes to female founders of technology businesses compared to their male counterparts at a global level and certainly at Africa level. In Africa specifically, the funding gap between male and female founded tech startups remains significant. Only about 2% of total funding goes to female-led ventures, while male-founded startups attract the vast majority of investment. For example, in the first half of 2023, female founded startups raised just $22 million, compared to the billions directed toward male-led ventures. This disparity has improved slightly but remains vast; female-led startups received 36 times less funding than their male counterparts during that period (note 1).

Several factors contribute to this gap. Female founders often lack access to key networks dominated by males and are less represented in high-investment STEM (Science, Technology, Engineering, and Mathematics) sectors. There is also a persistent gender bias in investment circles, with female-led ventures sometimes viewed as riskier despite evidence showing that they perform just as well as male-led ones.

However, if we were to take a step back, we would probably see that this funding ratio is positively correlated to the number of female founders – there just aren't that many in this space. I recently read a report by Startup Genome on the results of a survey of over 5,000 startups across 67 global ecosystems which ran between September 2016 and November 2022. The results showed that 31% of the respondents reported having at least one female founder, with the overall average of female founders in an ecosystem (total number of female founders divided by the number of startup founders surveyed) coming to just 15%. Clearly, there's a direct correlation between low participation and the low funding female founded business access.

As an angel investor (a person who commits patient capital to early-stage businesses) and someone who has driven several collaborations in the corporate world with technology leveraged businesses, I have often asked myself:

"Why?"

 DOI: 10.1201/9781003540618-11

"Is it that there aren't enough female founders or the ones that are available aren't necessarily investable?"

In my experience, the answer is a mixture of the two – sort of a catch 22 if you will. There's little incentive for women to show up if they don't see a sizable number of other women getting funding/partnerships, etc. And because few show up, the numbers remain low, and the repetitive cycle continues. I came across a 2022 International Women's Day article by Enrich Africa, where they interviewed Ruth Iselema, the CEO and founder of the crypto platform, and she also noted this perspective in the article where, according to the interviewer, "she revealed that many of the people who inspire her in the tech space are men because 'there aren't that many women in the tech space to look up to.'"

In addition to this of the few that do show up, there's often still work to be done to get both the founder and the business to the level of being investable, with limited avenues available for closing that gap. To bridge this gap, a multipronged approach is needed, including increasing female representation in investment circles, offering mentorship, and creating more female-focused funding initiatives. In this chapter, I will focus on mentorship as an approach to unlock significant value and make a discernible difference for women-led businesses, especially those in technology.

A Business Mentor Can Add Value

I recall a business pitch session where a few startups were put in front of potential investors and allocated the same amount of time to pitch their business ideas. There were five pitches in all, with a fairly even spread in gender representation among the overall team members. However, when it came to pitching, there were two females compared to three male presenters. I observed something quite interesting – the men generally spoke in bold terms about their businesses and the potential opportunities ahead for returns. The women, on the other hand, were almost apologetic about their business and pitched as though investors were doing them a favor, even though in terms of product definition and execution, they offered a far superior opportunity. This meant that they pushed for lower values of investments and didn't exude as much confidence as their male counterparts.

Sadly, these were not isolated incidents, and I have seen this scenario repeated over the years in my role as an angel investor and participating as a judge in multiple varied pitch events, from local universities to international platforms such as Jack Ma's Africa Business Heroes. The same experience has been corroborated by other investors I know. This spoke clearly to me that there's a need to provide business mentorship training to female founders – not necessarily on the technical aspects of the business, which they generally seem to have a handle on. Rather, the focus would be on the ability to demonstrate business leadership and the capacity to position their businesses as worthwhile investments.

These experiences led me to do some digging to try and understand how this situation came to be in the first place. One of the first areas I looked at was the sociocultural element, and it became clear that while most economies acknowledge the need for female entrepreneurship, not many actively encourage women to take up economically prominent roles. This is probably why we see the range of female founded businesses in North America and Europe skewed at a much higher level than those in Africa (Mastercard Index of Women Entrepreneurs 2021 report).

Some of the critical elements that pushed forward women's ability to be economically prominent beyond the overall wealth quotient of the economy and government influence include the seemingly "softer" elements, which include:

1. A positive sociocultural perception of female entrepreneurship.
2. Inspiration to make a difference in the world.
3. Fairly strong perception of capabilities.

When we look at female founded businesses in the technology space, these three factors create a solid base from which strong female founders can emerge or be hindered if they are absent. This is where mentorship can make a difference.

WHAT MENTORSHIP IS AND ISN'T

Mentorship provides critical support in navigating obstacles women face in the tech space, which include, but are by no means limited to, persistent gender disparities, limited access to capital, networks, and industry expertise. The right mentorship platforms create avenues for guidance, knowledge transfer, and access to influential networks that can be pivotal for both personal development and business growth. Having the right mentor(s) can dramatically alter the trajectory of a female founder. This makes it critical to define what mentorship does and does not entail.

Mentorship is a professional relationship in which a more experienced individual (the mentor) provides guidance, support, and advice to a less experienced person (the mentee). The goal of mentorship is to help the mentee develop the necessary skills required to navigate challenges and achieve personal or professional goals. Mentorship can be formal, with structured programs, or informal, where the relationship evolves naturally over time between the mentor and mentee. For those seeking to be mentors or seeking mentorship, defining the key aspects of a mentorship relationship in advance is vital to the success of the relationship so expectations are satisfactorily managed. There are four key things to expect in a mentorship relationship:

- **Guidance and support**: Mentors are expected to share their knowledge and experiences to help mentees grow. Mentors are not counselors or lecturers, and advice given should be within the context of what the mentee requires, rather than a constant download of one's life experience. If, as a mentor, you are not open to doing this, then it may be best to let the mentee know upfront that a mentorship relationship will not be possible.
- **Skill development**: The role of a mentor isn't that of a parent or teacher. As such, mentors should guide rather than prescribe how a mentee improves specific skills or learns new ones. The mentor has the benefit of years of experience and access to a breadth of information that the mentee can leverage to make progress in the identified areas of deficiency.

- **Networking**: Mentors should be open to introducing mentees to their professional networks and, where possible, creating new growth opportunities. However, the mentee should know upfront that the mentor isn't obliged to do so, and neither is the mentor responsible for client acquisition or fundraising. I say this because I have encountered situations where the mentee had an expectation that the mentor should take on fundraising responsibilities, sometimes injecting their own funds into the business. The mentor is free to do so, but it's not a requirement. And in the event that the mentor becomes a shareholder then the terms must be laid out clearly in advance in a legal document.
- **Confidence building**: A mentorship relationship has the ability to boost the mentee's confidence by consistently encouraging and empowering mentees to take on new challenges and pursue their goals. This is particularly important when the mentee has experienced setbacks or failures in business. If you have ever observed a pitch session, it will be clear that the confidence the founder exudes is a powerful determinant of whether the business receives funding or whatever support is required at the time.

Due to social and cultural influences, women tend to take rejection more personally than men, even in terms of investments. This means hearing "no" from an investor may be taken more personally by a woman than her male counterpart, which can impact future pitches negatively. Confidence building is also necessary because women may not be as assertive as their male counterparts, not just in pitching but in overall business leadership.

MENTORSHIP OR SPONSORSHIP?

> *A mentor is a person who gives a protégé advice, answers questions, strategizes career moves and professional development. A sponsor is someone who promotes a protégé to other people to help advance the protégé's career. A pithy way to remember the distinction is this: "Mentors talk to you. Sponsors talk about you to powerful people when you aren't in the room."*
>
> **Simon K. Johnston**

> *A person can only truly be a sponsor if they are able to create an opportunity for you that you otherwise would not be able to access on your own. Your sponsor not only has more experience than you in a certain area, but also has the ability to bring you "up" alongside them.*
>
> **Chantal Brine**

"Mentors help you skill up. Sponsors help you move up." Jo Miller

The key difference between mentorship and sponsorship lies in the nature of the relationship and the roles each plays in career or business advancement. We have established that mentors focus on providing guidance, direction, networking opportunities, and confidence boosting. Sponsors, on the other hand, are focused on directly and actively influencing the growth of the individual or business.

Sponsors are advocates who intentionally use their power, influence, and network to open doors for the protégé, offering opportunities for promotions, high-visibility projects, or critical connections. A sponsor puts his/her reputation on the line for the protégé and is therefore invested in actively engineering a successful outcome. There's nothing passive about sponsorship, and it involves the deployment of a lot of social, political, and sometimes financial capital to achieve the desired outcome.

I recall reading about the former prime minister of the United Kingdom, Margaret Thatcher's career history as part of preparation for a board readiness program, and it struck me that the concepts of mentorship and sponsorship clearly played out in her career advancement. When Margaret Thatcher became active in politics, Iain Macleod was the shadow chancellor and her boss, who acted as a mentor by taking her under his wing and guiding her in understanding how to navigate the corridors of political office. However, it was Edward Heath who actively sponsored her career growth by consistently creating opportunities for her to take on more responsibility and gain an increasing amount of visibility to the point where she became not just a participant in the government, but a member of the cabinet. This exposure ultimately forged the platform from which she became, in 1979, the first female and, to date, the longest-serving prime minister of the United Kingdom.

I used the example of Margaret Thatcher because politics is quite similar to the technology space with respect to its nature of being largely male-dominated. Even when there is female presence, it tends to be more from a supporting rather than a leadership perspective. In politics, as with technology, the presence of powerful female role models is scarce, and systemic barriers are more pronounced.

Mentorship or Accelerator programs?

In the world of technology businesses, we often hear of "accelerator programs" which are designed to fast-track the business growth of an enterprise. As with mentorship, accelerator programs also offer guidance and support, but they differ in structure, duration, and outcomes.

While a mentorship relationship can be long-term, formal, or informal depending on the agreement between the mentor and mentee, accelerator programs are structured, short-term programs aimed at rapidly growing a startup or early-stage company. Accelerators are highly organized and intensive, typically lasting a few months. Accelerators have a clearly defined structure and expected outcomes as startups typically join cohorts, receive mentorship from industry leaders, access resources such as office space, and often get initial seed funding in exchange for equity. This is because accelerators are focused on scaling businesses quickly, preparing startups for investment, and providing access to a network of investors and advisers. The end of the program is signaled by what is called a "demo day" where startups pitch their businesses to potential investors. In essence, while mentorship provides ongoing personal and professional guidance, accelerator programs are short-term, intensive, and designed to fast-track business growth.

This level of detail is being shared so female founders can identify what type of program or boost they require at each point, as mentorship, sponsorship, and accelerators all have a role to play in driving the success of a business.

Is Mentorship Right for You?

There are clear benefits to being a mentor as it provides an opportunity to directly and positively impact the life and business of the mentee. For the mentee, connecting with the right mentor creates a framework for having an experienced person contribute to their personal development, enhancing their leadership skills and fostering resilience. The feedback loop and psychological safety mentorship helps mentees avoid common pitfalls and leverage opportunities effectively. However, it is critical to note that not everyone can mentor or be mentored effectively.

Here are four things to look out for in your mentorship engagement:

- **Mentorship agreement**: There must be a written mentorship agreement which covers, at a minimum the mentee's expectations, the frequency and platform for engagement, and the tenor of the mentorship relationship. This does not need to be a legalistic document, but it is important to have this upfront so both parties are aware of their respective expectations. Things may start off fairly fluid, but expectations must be clearly established and communicated.
- **Willingness to change**: Not everyone is open to change, especially when it comes to their business practices. When mentoring a founder, it's important that the mentor appreciates that the business is almost like the founder's baby, as it were. As such, it may be difficult for the mentee to be completely objective about gaps in the product or perhaps even gaps in themselves. If the mentee is not willing to acknowledge those gaps exist, then there is very little the mentor can do to make any form of impact.

I recall engaging with a founder who had sought out a mentorship relationship but was quite defensive about any suggestions made regarding the approach to the business, minimum viable product, sourcing funding, etc. We eventually got to the point where we came to an agreement that what the person wanted was not a mentor, but a financial sponsor for the business as it was. I was not open to this option at the time, given the observations I had about the business, and we parted ways on a positive note.

- **Commitment***:* The level of value the mentee places on your input is often evident in their approach toward your agreed meeting times and vice versa. Ideally, you both have an agreed frequency for your mentorship sessions, and if you or the mentee consistently miss these appointments, then it's important to re-evaluate and decide what needs to change. Perhaps, if the time selected no longer works for both of you, you need to come up with another time. If the lack of commitment comes from the mentee consistently not giving feedback required as and when due, then you must address whether the relationship is actually important to him/her.

As a mentor, you must also be clear that you will keep the time commitments made and give the mentee your full and undivided attention at the meeting times. If you know in advance that your schedule does not allow for this, then it is best to let the mentee know upfront. Remember, the mentee is also busy trying to run a business, so any time spent with you as the mentor must be productive and constructive.

- **Mentorship focus**: As I shared earlier, there is a difference between mentorship, sponsorship, and investing. As a mentor, you are not obliged to invest in the mentee's business or open your personal network to the mentee. Likewise, as a mentee, such demands should not be made of the mentor, but where applicable, they can be welcomed if volunteered. If the focus is on client acquisition, network access, etc., please be upfront about it. The mentor has the right to say yes or no, but at least the expectations are shared upfront.
- **Connection**: I recall someone I did not know reaching out to me on LinkedIn, and in the message requesting a connection, the person said the reason for reaching out was to, "make use of my network." While I appreciated the person's honesty, it baffled me somewhat that someone I did not know personally or professionally would think it was appropriate to have such an expectation. I declined the connection because accepting would mean I was open to the option, and I had no clue who this person wanting to come into my space was.

The point I am emphasizing here is that it's important to establish a connection before making demands of a person's time, expertise, network, etc. The nature of the business world today means that a lot of initial connections may be virtual, making it vital to introduce yourself and your business first to a potential mentor (ideally with links to your online presence), stating how you came to hear about the person and that you would like to connect with them. Requests should only be made after establishing a connection and the other party has welcomed you into his/her space.

Benefits to the Mentor

The benefits of mentorship should not flow in only one direction, as there is mutual value to be derived in these relationships. We have mentioned the benefits mentees gain, but there are upsides for mentors as well for contributing to the growth of another individual/business. Some of these benefits include:

- **Alternative perspectives**: The mentee often provides the mentor with a different perspective about the business, industry, opportunities, etc. that the mentor may not have considered. By mentoring another person, the mentor can essentially see life through the eyes of another person.

- **Leadership development**: Being able to successfully mentor someone adds to the mentor's leadership development capabilities as the mentor learns how to communicate experience, skills, etc. to the mentee in a manner that the mentee is able to absorb and apply to their situation.
- **Giving back**: Having garnered a wealth of experience over the years in one or multiple fields of expertise, as a mentor you can give back to your community, sector, and so on by intentionally partnering with existing mentorship platforms or creating your own. I would recommend partnering first, as established programs tend to have laid down structures, procedures, and existing mentorship networks you can learn from, rather than trying to start afresh on your own.
- **Relevance**: By actively mentoring leaders, especially in the technology space, mentors have the opportunity to stay abreast of new disruptions in the space, which can be applied in their own organizational or personal spaces.

Success Stories: Rising Tide Africa

We have established that mentors can help female tech founders build confidence, develop strategic business acumen, and overcome implicit biases that may affect their professional journey. How then do these founders find the right mentors who understand their unique challenges and can add the required value? Typically, these occur through referrals and can be sporadic in nature, making it challenging to track the impact of mentorship on a sector.

This suggests to me that while informal, one-on-one programs are beneficial, there is value in intentionally creating mentorship platforms that female founders can access and benefit from.

Over the last two decades, I have had the privilege of being a career mentor in formal programs such as Women in Business and Politics (WimBiz), Women in Successful Careers (WISCAR), and Rising Tide Africa (RTA – a group of female angel investors focused on investing patient capital in female founded/led technology-based businesses). I have also had the privilege of mentoring female career professionals and business leaders one-on-one outside of a formal program structure. As such, I have had the benefit of seeing mentorship in action and the positive impact it has had on the lives and businesses of the mentees. Given that the focus of this book is on women in technology, I will draw from the RTA experience to illustrate my perspective.

Founded in 2016, RTA is a network consisting of over 100 female investors who are leaders in their respective fields. RTA aims to bridge the gender financing and inclusivity gap in the investment, entrepreneurship, and technology ecosystem, striving to be an impact driver for women-led businesses in Africa, promoting innovation, equality, and sustainable development through its "MINE" pillars: Mentoring, Investing, Networking, and Education.

From its early days in Nigeria, RTA founding members quickly found that to effectively invest in female founded businesses in Africa, the immediate needs of women entrepreneurs went further than the focus on capital. The areas of mentorship and education had to take pole position in the investment approach due to the peculiarity of businesses on the continent of Africa. To address the need to close the female funding gap, RTA created a network-based platform to mobilize women angel investors and channel their capital and professional experiences toward empowering upcoming women-owned and -led businesses. This approach is particularly vital given the male domination of the technology field in Africa and the comparable dearth of female role models.

I joined RTA in 2020, and in addition to being an investor, I have also been an active participant in the mentorship and education pillars, which has given me some of the insight I shared in this chapter. The mentorship program focuses on empowering female founders through:

- Capacity building programs.
- One-on-one mentorship clinics between female founders and industry leaders.
- Facilitation of introductions to potential investors, partners, and customers, which are crucial for scaling a tech business.

As a result of following this approach, RTA has mentored over 200 female founders, invested ~$2M in over 34 companies across Africa to date, and remains steadfast in its commitment to supporting women entrepreneurs and fostering innovation in Africa.

The benefit of such structured programs is that the impact can clearly be tracked in terms of the development of the founders, their businesses, and funding such businesses attract, as ultimately the goal is to reduce the funding gap. The market is certainly robust enough to accommodate many more of such mentorship programs, which means there is room for multiple collaborations.

In conclusion, mentoring female tech founders is essential to fostering a more inclusive and innovative tech industry. It empowers women to succeed, promotes gender diversity, and ultimately contributes to a more robust and dynamic entrepreneurial ecosystem. By bridging the gap between potential and opportunity, mentorship ensures that talented female entrepreneurs can fully realize their business aspirations.

12 Building the Pipeline

Supporting Girls in STEM to Become Women in STEM

Oreoluwa Lesi

UNDERSTANDING THE GENDER GAP IN STEM

Across the world, there is a shortage of women participating in Science, Technology, Engineering, and Maths (STEM)-related fields. According to the United Nation's 2024 Global Gender Snapshot, globally women make up only 35% of all STEM degree graduates from higher institutions. Women are also under-represented in the STEM workforce. Although the World Economic Forum's 2024 Global Gender Gap report reveals a gradual rise in the number of women working in STEM jobs since 2016, at 28.2%, women are still very much in the minority. In addition, women are less likely to rise to leadership positions within the STEM workforce (12%) than in the non-STEM workforce (24%).

This lack of women leaders in STEM has major ripple effects for women's participation overall, from making STEM careers appear less attractive to younger women to male-centric work cultures and values that do not enable women to flourish. This leads to an under-representation of women and diverse voices at the leadership and policymaking levels.

The 2023 Future of Work report by the World Economic Forum shows that technology adoption – particularly of technologies such as artificial intelligence and automation – is expected to radically transform businesses over the next five years. This means that people who lack the requisite skills and qualifications to help shape the development and use of these systems will be disadvantaged.

As digitalisation increasingly railroads across different aspects of our lives, technology and other STEM-based innovations' power to enable and amplify existing contexts for good and for bad also increases. This means that inequalities that already exist between men and women's access to and benefits from certain rights – such as education, economic opportunities and decent work, healthcare and food – will be further heightened.

There are different issues at play here. Firstly, women are not pursuing STEM degrees in large enough numbers. Secondly, women who start on a STEM career path drop off, and very few make it to top leadership positions. Thirdly, the lack of female representation affects how accessible and attractive STEM careers are to girls and young women. This, in turn, affects the degree of encouragement given to girls

DOI: 10.1201/9781003540618-12

and young women to pursue what are now regarded as masculine career options by parents, teachers and other influencers.

Many STEM career opportunities depend on having a STEM-related tertiary-level degree or qualification. This typically requires the completion of secondary school and having studied science, technology and maths subjects.

According to the Global Snapshot 2024, in sub-Saharan Africa, "upper-secondary out-of-school rates were 48 per cent for girls compared to 44 per cent for boys." Girls from the poorest households were least likely to attend school at any educational level.

THE NIGERIAN STORY

Nigeria is one of three countries with the highest number of out-of-school children in the world. In 2020, there were 20 million children not enrolled in school in Nigeria (International Centre for Investigative Reporting, 2023). More than 50% of Nigerian girls of school age do not attend school at the basic education level. According to UNICEF Nigeria (2022), "1 million girls drop out between the first and last year of primary school, while 0.6 million leave between Primary 6 and Junior Secondary School 1." Like a funnel, a decreasing number of girls progress through the education levels from primary to tertiary.

In both the global and Nigerian cases, where young women can attend university or polytechnic, the enrollment of women in STEM degrees is overall lower than men's. Although some subjects – such as medicine, mathematics, biological and agricultural sciences, psychology and social sciences – have trended upwards, other subjects such as engineering and computer science have remained low. In Nigeria, only 22% of the Engineering and Technology university graduates each year are women (Vanguard Newspaper, 2023).

CAUSES OF AND STRATEGIES TO BRIDGE THE GENDER GAP

So, why aren't women choosing to study STEM careers? There are many reasons that have been posited over the years.

For a start, students' career choices are largely dependent on how well they perceive themselves to be good at these subjects in secondary school. So it is students who feel confident about their maths and science abilities; who tend to choose to study these courses in secondary school and university. Girls have been shown to assess their abilities more critically than boys and are less likely to perceive themselves as good at maths and science. As a result, they disengage from these subjects earlier in secondary school and are less likely to think about pursuing related careers.

It also appears that having social support is important to girls when deciding on a future career. If girls believe that they will not get the necessary support and encouragement from their family (Meyer, Cimpian, & Leslie, 2015), they are less likely to choose to study these subjects. Ekine (2016) noted the important role played by teachers. Girls' and young women's sense of identification with science and maths can be eroded by teachers who call on girls less than boys in the classroom or who, in other ways, perpetuate stereotypes of boys being naturally more suited to science and maths than girls. If there are fewer female science or maths teachers beyond the primary school level, this perpetuates the belief that science is not a suitable subject for girls. This is compounded by science textbooks, which frequently feature gender stereotypes and can also negatively affect girls' interest in the subject.

Research on diverse populations in STEM, such as girls and women of colour, by Ong, Smith, and Ko (2018) showed that "safe social spaces," which could provide women with support in STEM, helped them feel a sense of belonging to the wider STEM community.

In addition to increasing the students' interest in STEM and perceptions of their ability to do these subjects well, other ways to increase female interest in STEM careers according to Schilling and Pinnell (2019), include highlighting the interpersonal aspects of STEM careers by showcasing the creative aspects and the impact on human welfare. They also found that designing curricula that incorporate problem-based learning has helped improve the students' motivation towards learning, enabled them to self-direct their learning and allowed participants to create meaning for themselves.

In response to the deficiencies within the formal primary and secondary education levels and the other factors identified above that have led to girls' limited participation in science and maths, many programmes have been designed targeted at girls and young women within schools and outside of formal education, to get them excited about science and maths subjects and encourage them to pursue related courses of study at the tertiary level.

For the rest of this chapter, we will focus on the informal (i.e., non-school) avenues of science and maths learning.

ABOUT INFORMAL LEARNING

While measures to improve STEM learning capacity among young people typically support formal in-school strategies, such as enhancing science and maths curricula and teacher training, the informal avenues of STEM learning are those that lie outside the formal school system and are available to everyone irrespective of age and background. Informal learning leverages everyday activities and settings, such as conversations in the home, visits to museums, television shows and activities like farming (National Research Council. 2009).

The Committee on Learning Science in Informal Environments was set up to examine how non-school settings could facilitate science learning in the United States. The committee assessed how learners of different ages could learn science across varied settings and spans of time. The committee also identified ways in which informal and formal science learning were similar and different.

Due to the symbiotic relationship between maths and science and because of how both subjects feed into technology and engineering knowledge and practice, in this chapter, we will apply the findings about informal science learning to the STEM disciplines and use the term STEM interchangeably with science and maths.

The committee identified four major ways in which science and maths learning happen in informal environments:

1. Everyday experiences – like walking in the neighbourhood, watching a sunrise or stars at night – can teach people about the natural world and help them develop important skills for science and maths learning.
2. Designed settings – like a science museum, zoo, aquarium, nature reserve, conservation centre – provide places where people can learn about natural phenomena and the world around them.
3. Programmes – such as structured non-school STEM programmes like camps, weekend or after-school clubs and holiday programmes, delivered by organisations like STEM-focused organisations, community centres and public libraries can stimulate the interest of adults and children in science and may result in them seeing a future for themselves studying or working in STEM.
4. STEM media – such as radio programmes, television shows, websites and other resources on the Internet and apps – which make science more accessible to people for STEM learning.

Across these four ways in which STEM learning happens, learning is driven by people who decide whether to participate in these learning experiences or not. There is little pressure to do well in them, and as a result, informal learning environments can be a safe environment for learning about science.

HOW STEM LEARNING OCCURS IN INFORMAL SETTINGS

The Committee on Learning Science in Informal Environments proposed a "strands of science learning" framework that identifies certain capabilities that support science learning within informal environments (National Research Council, 2007).

Strand 1: Be excited about and want to learn about phenomena in the natural and physical world.
Strand 2: Understand and use scientific principles, explanations and facts related to science.
Strand 3: Test, question and make sense of the natural and physical world.
Strand 4: Reflect on science as a way of learning about phenomena in the natural and physical world.
Strand 5: Participate in scientific activities and learn with others, using scientific language and tools.
Strand 6: Identify themselves as science learners who know about and sometimes contribute to science.

In STEM camps and after-school programmes, we can see all these strands of learning happening in succession or sometimes simultaneously. The curriculum and activities are designed to engage and educate students on STEM concepts. They are also intended to give students the experience of solving problems using STEM principles and to influence their attitudes about and sense of identification with STEM.

Camps vary in terms of subject focus, age, length and intensity, but typically, the sessions are led by a trained facilitator and can include classroom-type learning sessions, hands-on team-based project-focused experiences, excursions or visits by industry professionals.

Assessing the outcomes of camps is not straightforward because there are different types of camps, participation is consensual and although there is the expectation for STEM learning, a lot of emphasis is also placed on having fun and making friends.

Researchers and implementers of informal STEM learning may struggle with the theoretical and practical aspects of measuring if and how learning occurs. While schools use standardised examinations or tests, reports or essays to measure how well students have mastered the concepts taught, informal STEM settings do not usually use assessment instruments, such as test scores or class rankings, because this type of standardised testing is not aligned with the informal learning types of activities and reasons for participating.

In addition, because it is difficult to determine which experiences and sessions led to which specific outcomes, to arrange for pre- and post-tests, or to control for influences, typical assessment practices in formal environments don't translate well.

Informal settings will usually use any combination of evaluation methods, including questionnaires, interviews, focus group discussions, thinking-aloud techniques, participant observation and video and audio recordings to gather data. The metrics measured could include self-reported data or interviews on attitudes, interests and beliefs, observations of learners' behaviour when engaging with STEM (National Research Council, 2009).

Since informal learning activities are not always implemented in a structured and controlled way, the outcomes may be in line with the expected goals or completely unanticipated. They may manifest at different points in time or have varying degrees of reach, i.e., at the level of the individual, the group or the community.

On the whole, according to the findings of the National Research Council (2009), in the United States, learning science in informal environments has the potential to advance science education across education levels and forms one of three key pieces of the national education system – along with K12 (Kindergarten to year 12) education and higher education – by "producing citizens literate in STEM concepts and future scientists, engineers, mathematicians, and technologists" (National Research Council, 2009).

STEM CAMPS FOR GIRLS

Although there isn't one clear way to evaluate how effective informal learning is, available evidence indicates that it can help present STEM in exciting ways, provide low-risk opportunities to test and apply science and maths concepts and influence students' perceptions about STEM.

One notable example is the University of Dayton. From as early as 1973, the University of Dayton in the United States of America has run a Women in Engineering Camp (Updyke, 2018). The one-week, residential summer camp is organised for high school female students on the university campus to give them some exposure to engineering careers. The women participate in hands-on sessions like "Experiments in Engineering" and "Innovation Modules." They are also able to meet and speak with practising engineers at the "Dinner with an Engineer" and a "Day with Industry."

Between 1973 and 2018, when the camp celebrated its 45th anniversary, the University of Dayton's School of Engineering female population had increased from less than 1% women to 24%, which is a significant increase.

Research shows that the ages of 11 to 15 years are a formative period for students' self-perceptions in mathematics and science, and where they decide whether to pursue a STEM path (Microsoft, 2017). After this, it is increasingly difficult to interest students in the STEM field. This means that programmes that target young women who are 15 years or older – like the University of Dayton Women in Engineering Camp – might be attracting young women who are already interested in STEM subjects and careers and missing the opportunity to create positive STEM perceptions for other young women whose interests or career aspirations are leaning away from STEM.

UNESCO (2017) also found that the supportive environments and introduction to real-life learning opportunities provided through extra-curricular activities, field trips, camps and internships, can help spark and keep girls' interest by increasing their self-confidence in STEM. This provides strong support for the "catch them young" movement, where informal learning targets girls from as young as 11 years.

In addition to targeting girls early, it also appears that girls-only programmes might be slightly more successful in encouraging girls to see a place for themselves in science and maths than co-educational programmes. Schilling and Pinnell (2019) tracked participants of three types of camps over two years:

- Female participants in a women's-only engineering camp (Group 1).
- Male participants from a co-educational (or co-ed) engineering camp, i.e., a camp with both women and men (Group 2), and
- Female participants from the same co-ed engineering camp (Group 3).

All camps lasted for six days and introduced the participants to a variety of engineering fields, such as civil, chemical, computer, mechanical and electrical engineering. In the first year, the participants of the female-only camp benefited from an industry day, where they travelled to industry partners to tour the companies or plants and learn more about engineering careers directly from engineering professionals. The co-ed camp did not include this type of exposure to engineering professionals but instead offered a robotics module, which ended with a competition where robots designed, built and programmed by the participants had to perform certain tasks.

Overall, the results for the first year showed an increase in the female participants' sense of confidence in engineering in both the single-sex and co-ed camps. However, responses to questions about how they viewed engineering and how they rated their competency showed lower scores for the female students from the co-ed camp than from the female-only camp. This suggests that the participants from the co-ed camp had experiences that decreased their sense of confidence about engineering.

It appeared that there were limited opportunities for the female participants to completely immerse themselves in the co-ed camp activities and workshops. For instance, the researchers observed that the female participants were frequently assigned the role of note-taker for their group, which restricted how much they could contribute to the group work. On another occasion, a complete explanation of the module was not given to the class because most of the male participants already had experience with the activity. This left the female participants (most of whom had no experience with the activity) unable to contribute fully.

It emerged that arts was a favourite subject of more female participants than male participants, but there were not many creative activities in the co-ed camp. So, in the following year, the camp organisers created more opportunities in the co-ed camp for participants to engage in creative activities that were linked to engineering, such as using the university's makerspace, exploring creativity in coding and learning about some newer fields like nanotechnology.

The camp organisers used the feedback and observations from the first year's camps to create more opportunities in both female-only and co-ed camps for the students to engage in creative activities, which allowed them to relate engineering to concepts beyond just maths and science. The camp organisers also ensured that the facilitators were more intentional about creating spaces where the female participants felt comfortable. As a result, they saw higher positive responses all around.

The conclusion of their research was that the female-only engineering camp model can have more positive effects on young women's beliefs in their ability to perform well in STEM.

There is other research that shows that female-only camps provide settings that enable girls and young women to gain more appreciation for STEM and for their self-efficacy in these subjects. Sultan, Axell and Hallstrom (2023) shared how girls and young women in co-educational environments may temper their performance in the presence of boys. While Phelan, Harding and Harper-Leatherman's study, as cited in Schilling and Pinnell (2019) demonstrated that STEM camps can be successful in increasing the interest of underserved populations – such as women of colour – in STEM.

However, Schilling and Pinnel (2019) observe that if a co-ed camp is well-designed, it can also help raise young women's sense of confidence with STEM.

Conrad et al. (as cited in Sultan, Axell, and Hallstrom, 2023) learned from their study of male and female students participating in a STEM summer camp that middle school students (ages 9 to 13 years in the United States) left the camp with more positive feelings about STEM than the elementary (5 to 8 years) and secondary students (14 to 17 years).

Overall, based on much of the research, it would appear that girls-only camps that target girls from the ages of 9 to 15 years and comprise of a carefully designed combination of workshops, hands-on activities, exposure to STEM in action via excursions and talks by industry professionals, internships or volunteering experiences might be one of the best ways to engage girls with science and maths subjects, help them develop opportunities to apply these subjects in meaningful ways and enable them to see a place for themselves in STEM.

SPOTLIGHTS OF SOME NIGERIAN GIRLS-ONLY STEM CAMPS

In Nigeria, there are dozens of STEM camps organised every year, although relatively few are targeted solely at girls. We searched for STEM camps that fit the following criteria:

- Girls-only.
- Target girls aged 11 to 17 years.
- Have taken place within the last two years, i.e., from 2022 onwards.
- Engage the girls for at least 30 contact hours within a one-week period.
- Based in Nigeria.

Based on these criteria, we identified Visiola Foundation's STEM Summer Camp and the Women's Technology Empowerment Centre (W.TEC)'s She Creates STEM Camp.

THE VISIOLA FOUNDATION STEM CAMP

The Visiola Foundation is a not-for-profit organisation established in 2014 and based in Abuja (the northern part of Nigeria) that invests in African girls and young women by providing STEM education, training and mentorship. In addition to STEM camps, the Visiola Foundation runs after-school STEM Clubs for girls in government secondary schools, coding boot camps for young women from across Africa and has directly impacted over 12,400 girls from 12 African countries since 2014.

The Visiola Foundation STEM Camp is a residential eight-day programme aimed at getting girls interested in the STEM fields.

During the STEM camp, participants are introduced to concepts in robotics. They explore how the practical applications of mathematics, physics and other science subjects work in the real world. They are taught to understand how science and technology can affect society and can catalyse the sustainable development of the country. They work both on individual projects and group assignments to develop their critical thinking and problem-solving abilities.

It is Visiola Foundation's intention to encourage more girls to pursue STEM careers by grabbing their interest early, building their confidence and encouraging their success. Through the camp, Visiola Foundation seeks to increase the number of girls in STEM with a view to creating a pipeline of professionals who will leverage technology in developing innovative solutions to Nigeria's most pressing challenges.

The STEM summer camps end with a closing competition where students come up with designs and models to solve problems for homes, factories, hospitals, public facilities and other applicable places for their final projects.

The camp is free, but students are required to pay a 1,000 Naira application fee, complete a 250-word statement explaining their interest in the camp and submit their report cards for the last academic year. Aside from the payment of the application fee, to qualify for the camp, the girls must:

- Be an African girl aged 13–18 years.
- Having completed her primary school education in any African country.
- Be currently enrolled in a secondary school in an African country.
- Be available to stay for the full duration of the camp.

The 2024 STEM camp had 100 girls in attendance. The impact report from the Foundation's website shows that 80% of the students who participate in the STEM summer camps and after-school clubs complete secondary school. Ninety percent of all students reached through all their programmes demonstrated improved academic performance, especially in maths and science subjects.

THE WOMEN'S TECHNOLOGY EMPOWERMENT CENTRE (W.TEC) SHE CREATES CAMP

The Women's Technology Empowerment Centre (W.TEC) is a Nigerian nonprofit organisation founded in 2008. It was one of the earliest Nigerian organisations working to address the gender gap in STEM. W.TEC is headquartered in Lagos (the southern part of Nigeria), although its work extends to many states across Nigeria. Over the last 16 years, W.TEC's programmes have encouraged over 48,000 girls to consider and pursue STEM careers and supported women to adeptly leverage technology to enhance their socio-economic power. W.TEC does this through programmes such as the after-school technology clubs for girls in Nigerian public secondary schools across three states and an annual residential STEAM camp.

W.TEC piloted the camp in 2008 with a focus on technology but gradually expanded in scope to include science, engineering, maths and arts.

The She Creates Camp is an intensive residential STEM-centred programme that targets girls aged 13 to 17 years. The camp is designed to engage girls with STEM at a fairly young age by providing information and organising fun, hands-on training to challenge stereotypes about who should be pursuing science, technology and maths-related subjects and careers. The camps have taken place in Lagos, Oyo, Kwara and Anambra states. The She Creates Camp provides a high-quality curriculum (including programming, mobile application development, graphics designing, film-making and digital animation programming) and a fun environment and over 680 girls have learnt to create or innovate with technology to solve problems in their everyday lives. By immersing them in project-based classes that solve real-life problems, exposing them to industry professionals and visiting STEM-based companies, the She Creates Camp has encouraged girls to develop an early interest in science, engineering, computer science, information technology and related careers.

The pilot edition of the She Creates Camp was focused solely on technology but gradually increased in scope to STEM and then to STEAM. This evolution was based on feedback from the girls and observations of the interdependent nature of the STEM subjects and how creativity was increasingly valued as an important skill for future jobs.

The camp is open to girls who meet the following criteria:

- Fall within the age bracket of 13–17 years
- Demonstrate a strong commitment to expanding their knowledge of technology.
- Are available for the entire period of the camp.
- The application process is categorised into scholarship applications and general applications. Both processes start off with the participants submitting the form and indicating if they need a scholarship. Scholarship applications are assessed based on the submission of the last term's report card and an essay explaining why they want to attend the camp.

The 2024 edition of the camp accepted 19 girls for the two-week residential camp. The She Creates Camp cost 250,000 Naira per girl; however, scholarships from sponsors enabled over half of the participants to attend the camp at no cost to them.

The theme for the 2024 edition of the camp programme was "Environmental Sustainability with Tech" and featured sessions on environmental sustainability, artificial intelligence, marine engineering, recycling and solar power. This was complemented by sessions on gender and leadership, public speaking, emotional intelligence, etiquette, financial literacy and digital safety.

The evaluation of the She Creates Camps showed that 96.2% of the participants have been using the skills and knowledge gained from the camp in their daily lives. 36.7% of the university-aged alumnae are studying STEM degrees at university, while 78.4% of the alumnae still in secondary school expressed their desire to study a STEM degree in university. Sixty two percent of the participants indicated that they are generating income from the skills and knowledge acquired from the camp. Overall, 69.6% of the participants stated that participating in the She Creates Camp motivated them to pursue careers in STEM.

At the 2024 graduation ceremony, three alumnae who had returned to serve as volunteer instructors for the camp shared their testimonies. Two were currently pursuing computer science and data analytics career paths and credited their paths to the exposure and encouragement they received from the camp. The third alumna was working as a copywriter and considered herself at the intersection of technology and art. She loved writing and was considering a related career. However, the camp – particularly the session on blogging – introduced new possibilities that incorporated technology and innovation with art.

RESULTS

Both Visiola Foundation and W.TEC demonstrate increased interest in STEM careers among the participants who attended their camps and completed the surveys. The Visiola Foundation shows that most participants of their camp perform better in school and especially in science and maths subjects after completing the camp. W.TEC shows that a majority of the camp participants singled out the She Creates Camp as a primary motivation for wanting to pursue a STEM career.

What is unclear is how many of these young women were confident in science and maths prior to attending the camps and were already interested in STEM careers. Based on data from other researchers, such as Schilling and Pinnell (2017), STEM camps frequently attract girls who are already leaning towards these subjects.

To have a clearer picture of how effective the girls-only camps are in motivating girls towards STEM fields, we would need to ask in pre- and post-tests about the young women's career aspirations. We would also need to identify and include more girls-only camps in our research and analyse the findings to determine how significant these results are and if they can be used to determine causality or correlation.

We might want to establish a control group, which could include a co-educational STEM camp targeted at the same age range and/or a group of girls who have not participated in a STEM camp or other informal STEM activity.

CONCLUSION

For now, based on the available research and the outcomes of the two camps spotlighted here, participating in a girls-only STEM camp can be successful in showcasing science, technology, engineering and maths in relatable ways. The camps can provide safe learning spaces for the girls to experiment, make mistakes and grow their confidence while engaging with STEM. Incorporating art-related activities can make the camps more fun and connect to the girls' creative sides. It can also help to demonstrate the creative aspects of STEM careers, which include thinking outside the box to solve a problem.

The pathway from deciding to pursue a STEM career as a pre-teenager or teenager to becoming a practising professional is a long one, and there is no guarantee that attending a STEM camp will directly increase the number of female professionals. Although not every woman who practices as a STEM professional has a STEM degree, enjoying and being confident in science and maths subjects still appears to be an important factor for them to even consider a STEM career. Therefore, as parents, educators, researchers, policymakers or whatever roles we occupy, we can be deliberate about finding or creating informal learning opportunities that weave in science and maths in fun and casual ways that will hopefully make pursuing a STEM career seem that much more attainable.

REFERENCES

Alanna Petroff. (2017). The exact age when girls lose interest in science and math. CNN Business. Available online: https://money.cnn.com/2017/02/28/technology/girls-math-science-engineering/index.html (Last accessed October 12, 2024)

Ekine, A., & Abay, N. A. (2016). Enhancing girls' participation in science in Nigeria: A driver for national development and social equality. Available from: https://www.brookings.edu/articles/improving-learning-opportunities-and-outcomes-for-girls-in-africa/ (Last accessed October 6, 2024)

International Centre for Investigative Reporting, September 24, 2023 https://www.icirnigeria.org/over-7-5m-nigerian-girls-not-enrolled-in-school-unicef/ (Accessed 2 August 2024. No longer accessible)

National Research Council. (2009). *Learning Science in Informal Environments: People, Places, and Pursuits*. Washington, DC: The National Academies Press. Available online: https://nap.nationalacademies.org/catalog/12190/learning-science-in-informal-environments-people-places-and-pursuits (Last accessed October 11, 2024)

Meyer, M., Cimpian, A., & Leslie,S. (2015). Women are underrepresented in fields where success is believed to require brilliance. *Frontiers in Psychology*, 6, 1–12 Available online: https://www.researchgate.net/publication/276839885_Women_are_underrepresented_in_fields_where_success_is_believed_to_require_brilliance (Last accessed October 2, 2024)

National Research Council. (2007). *Taking Science to School: Learning and Teaching Science in Grades K-8*. Washington, DC: The National Academies Press. Available online: https://nap.nationalacademies.org/read/11625/chapter/1#iii (Last accessed October 11 2024)

Ong, M., Smith, J. M., & Ko, L. T. (2018). Counterspaces for women of color in STEM higher education: Marginal and central spaces for persistence and success. *Journal of Research in Science Teaching*, 55(2), 206–245. Available online: https://onlinelibrary.wiley.com/doi/10.1002/tea.21417 (Last accessed October 7, 2024)

Schilling, M., & Pinnell, M. (2019). The STEM gender gap: An evaluation of the efficacy of women in engineering camps. *Journal of STEM Education*, 20(1), 37–45. Available online: https://www.jstem.org/jstem/index.php/JSTEM/article/view/2346 (Last accessed October 9, 2024)

Sessou, Ebunoluwa. 2023. *APWEN Says Percentage of Women and Girls in STEM is Worrisome*. Vanguard Newspaper. Available online: https://www.vanguardngr.com/2023/02/apwen-says-percentage-of-women-girls-in-stem-worrisome/ (Last accessed October 9, 2024)

Sultan, U., Axell, A., & Hallstrom, J. (2023). Bringing girls and women into STEM?: Girls' technological activities and conceptions when participating in an all-girl technology camp. *International Journal of Technology and Design Education*, 34(2). Available online: https://www.researchgate.net/publication/373648959_Bringing_girls_and_women_into_STEM_Girls'_technological_activities_and_conceptions_when_participating_in_an_all-girl_technology_camp (Last accessed October 7, 2024)

United Nations Educational, Scientific and Cultural Organization (UNESCO). (2017). *Cracking the Code: Girls and Women's Education in Science Technology Engineering and Maths (STEM)*. Available from: https://unesdoc.unesco.org/ark:/48223/pf0000253479 (Last accessed October 9, 2024)

UNICEF Nigeria. (2022). Education opportunities for Girls in Nigeria. https://www.unicef.org/nigeria/media/7746/file/UNICEF%20Nigeria%20Cheat%20Sheet:%20Out-of-school%20Children.pdf (Last accessed October 9, 2024)

UN-Women and DESA. (2024). Progress on the Sustainable Development Goals: The Gender Snapshot 2024. New York: UN-Women and DESA. Available from https://www.weforum.org/publications/global-gender-gap-report-2024/ (Accessed on October 5 2024)

Updyke, K. (2018). Women in engineering summer camp marks 45 adventurous years! Available online: https://udayton.edu/blogs/engineering/2018/18-07-20-wie-dwe.php. (Last accessed October 12 2024)

World Economic Forum. 2024. Global Gender Gap 2024 Insight Report Available from: https://www3.weforum.org/docs/WEF_GGGR_2024.pdf (Accessed on 5 October 2024)

13 Inspiring the Next Generation of STEM Women

Opportunities and Strategies

Sandra Ilogu

INTRODUCTION

Women are underrepresented in Science, Technology, Engineering, and Mathematics (STEM) fields, especially in Information Technology (IT), where only a fraction of roles are occupied by women. Despite a 1.58% increase in STEM female workers between 2015 and 2023, a troubling gender imbalance remains with women comprising only 29.2% of the STEM workforce in 146 nations evaluated, compared to nearly 50% of non-STEM occupations.[1]

The extent of this gender disparity varies significantly by region and country due to cultural, political, and socioeconomic factors. Taking everything into account, the global underrepresentation of women in STEM is a concerning issue, and work must be done to remove any barriers that prevent women from starting, remaining within, and advancing in STEM careers.

This chapter explores the underrepresentation of women in the male-dominated space of Information Technology (IT), the need for a diverse and inclusive presence of women within this field, and incorporates my experiences and external studies to outline the various obstacles encountered and offer practical approaches to overcome these challenges. It aims to build the next generation of confident women, ensuring they thrive and create sustained success for the following set of women in the fray. As IT is part of STEM, there will be references to STEM in a broader context and thoughts on opportunities for further improvement within society, educational institutions, organizations, governments, and the scientific and technological community.

BACKGROUND

The path of women in IT started on a promising trajectory with great pioneering contributions, which fell with the onslaught of significant barriers. Now, it is on the rise again with great efforts to rebuild and increase representation.

DOI: 10.1201/9781003540618-13

Female pioneers include: Ada Lovelace (1840s), known for writing the first algorithm for Charles Babbage's Analytical Engine; Grace Hopper who contributed to the creation of COBOL and laid the groundwork for modern software development; and Katherine Johnson, Dorothy Vaughan, and Mary Jackson, the African American mathematicians and "human computers" at the National Aeronautics and Space Administration (NASA), who played a key role in the success of US space missions.

In the early days of computing (1940s–1960s), programing was often seen as a clerical or administrative task, akin to secretarial work, and thus viewed as "women's work." However, as the field grew in importance, particularly in the 1970s and beyond, programing and IT became more prestigious and were rebranded as highly technical, "masculine" work. This shift marginalized women and is documented in works such as Marie Hicks' "Programmed Inequality" (2017), which discusses how the professionalization of computing pushed women out of key roles in the UK as it became a prestigious male-dominated field. Hicks details how in post-war Britain, the computer workforce initially included many women, but as computers became more integrated into defense and high-prestige government projects, men were brought in to "professionalize" the field. Women were either pushed into less prominent roles or excluded entirely, and with the introduction of the home computer, this trend continued with boys dominating early exposure leading to a gender gap in early exposure to computers.[2] This disparity meant that by the time boys and girls got into higher education, boys were more familiar with technology, creating a self-reinforcing cycle of male dominance in computer science programs.

This shift reinforced the idea that men were more "naturally" adept at math, logic, and technology, while women were seen as less capable in these areas. The field was rebranded as a highly skilled, male-dominated profession, and the change was driven by a combination of cultural stereotypes, marketing, professionalization, and workplace practices that systematically marginalized women from roles they had once helped pioneer, resulting in the global underrepresentation of women in IT. This issue is of great concern for several reasons:

1. Loss of talent and innovation potential: STEM fields thrive on diverse perspectives, which lead to more innovative solutions and discoveries. When women are underrepresented, their unique insights and contributions are lost, limiting the overall potential for breakthroughs and advancements. A more diverse workforce fosters creativity and a broader range of ideas.
2. Economic impact: STEM careers are often among the highest-paying and fastest-growing fields. When women are underrepresented, they miss out on significant economic opportunities, contributing to a gender pay gap. Furthermore, industries miss out on the economic benefits of having a larger pool of highly skilled workers.
3. Gender inequality and stereotypes: The underrepresentation of women in STEM reinforces harmful gender stereotypes and the notion that certain fields are "pro-male" or unsuitable for women. This can discourage young women from pursuing STEM careers, creating a self-reinforcing cycle of exclusion.

4. Systemic inequality: In many parts of the world, social norms, cultural expectations, and institutional biases still discourage women from entering STEM fields. This limits their access to education, mentorship, and opportunities for career advancement.
5. Impact on research and development: The absence of women in STEM can skew research priorities and outcomes. For example, research in health, technology, and engineering may neglect the specific needs and perspectives of women, leading to products and solutions that do not adequately serve half the population. Companies like Johnson & Johnson have reported that a mixed-gender workforce in healthcare teams leads to more comprehensive approaches to patient care, as women and men often bring different insights into patient needs, health challenges, and innovations in medical technology.

Efforts to rebuild representation for women in IT are ongoing, as advocacy, educational reform, and diversity initiatives work to address these long-standing challenges.

MAIN CONTENT

My journey into IT was unconventional; it was not planned, but looking back at my beginnings, and the way things lined up, it is of little wonder that this is where I ended up. Like most children, I was very inquisitive, and I believe that my journey into IT was shaped by a collection of subliminal influences that fueled my innate curiosity and desire to understand how the world works. From a young age, I was fascinated by the inner workings of everyday objects and was naturally inclined to dismantle things like watches just to see how they functioned. This love for discovery and solving was not just a passing hobby; it was a core part of how I interacted with the world based on events in my life, which in hindsight nurtured my early signs of a budding scientist.

Listed are the influences that I believe propelled me in the STEM/IT space.

1. **Childhood activities**: I have one brother, and because there were just two of us, team games were the order of the day. Though I had dolls and he had his toy cars, a lot of times, we would get gifts that we could play with together. One such gift type was chemistry sets, which our mom would buy anytime she traveled. Receiving them represented an open door to fun and experiments with our cousins and friends. Mixing the chemicals to produce smelly gases, changing the color of liquids depending on acidity, or going around the house with my litmus paper to see what items were acidic or alkaline were fun activities. The experience of mixing chemicals in test tubes, observing reactions, and learning about scientific principles through play set the building blocks for my passion for science. I later saw science not just as a school subject but as a thrilling exploration into how things around me could be understood and manipulated.

2. **Television**: We had set TV times and there were shows that we loved to watch. One was *Space: 1999,* which exposed me to futuristic technology and the world of computers. I loved watching how the Alphans on Moonbase would press a few buttons and have printed information coming out of some machines in the walls. Watching the computer outputs on the show sparked my imagination about the possibilities of technology. It wasn't just entertainment; it was a glimpse into a future I wanted to be part of. Computers seemed magical and powerful, and I wanted to understand the logic behind them, how they processed information, and performed tasks that seemed impossible for human minds. Watching that show further validated my interest in technology, even before I knew what personal computers looked like.
3. **Academia**: I was placed into the default science class at school. This wasn't because I specifically chose to study science; rather, I was good at all subjects, including science and mathematics, so based on the institutional guidelines, that was the class I was placed in or belonged to. I was now exposed to formal scientific learning in a structured environment where I got to study other subjects like physics, chemistry, and biology.
4. **Peers**: I had a great group of friends with common goals of being the best we could be. We encouraged, advised, shared opportunities, supported, and held each other accountable as we meandered through adulthood. We all planned to go abroad for further studies, work, buy a car (and some of us thought of buying a flat too), and head back to Nigeria to work, get married, and start a family.
5. **Scholarship**: My first degree was in chemistry and as planned with my group of friends, we headed out to England after our Youth Service, in my case, for the new rave, an MBA. The school I wanted to go to was expensive, so I deferred my admission and started working with a plan to save toward the course there or at another good MBA school in Europe. Through my "Peers" group, I found out that there were universities in the north of the United Kingdom where Information Technology was being encouraged for women. They were prepared to sponsor females to complete a Master of Science in Computer-Based Information Systems—full tuition and board and there was promised assistance or employment on completion. It was a win-win. I took that on. I got admission and headed up to Sunderland—far away from my familiar territory of London to get this degree.
6. **Role models**: The fact that I had role models who had great careers in computing helped make the decision to switch from an MBA to the MSc an easy one. My late father was the Head of Computer Auditing for the Federal Ministry of Finance, and my late uncle was a Program Head at Schlumberger.

On completion of my Masters, I was a new female entrant in the IT field in the UK. I had to face the challenges that come with being a woman in this male-dominated space. My first "real" job was at a small insurance company as an Oracle programer in Oxford. I was the youngest person and only black female in their computing department. The other female employees provided administrative support and were in a different building. With just male counterparts surrounding me, the disparity in gender representation was glaring and presented a few challenges. My next move was to work at a large global corporation that had an IT application development group in which there was a higher number of women.

Challenges Encountered

1. **Perceived role:** In my early days at the insurance company, I was assumed to be present at meetings to take notes or ensure that whatever food we had for a meeting was adequate and to the liking of the meeting attendees.
2. **Imposter syndrome:** I felt a highly burdening need to prove myself. This meant working twice as hard as my male colleagues to show that I was efficient, smart, and a value deliverer to my team and the company.
3. **The tightrope walk:** There was a constant balancing act of doing the right thing, asking the right number of questions, challenging in the right way in order to earn and keep the same respect that many of my male colleagues received naturally and not come off as a "fusspot" or "whiney lady."
4. **The office etiquette** left a lot to be desired, and the foul language made the space psychologically unsafe. This time, however, was long before the current ways of working, so one had to have a "thick skin." The same applied to outside office activities, which was typically to go to the pub, get smashed and talk about the experience the next working day. It was lonely!
5. Even though there were more ladies in the global organization where I later worked at, there was **palpable hostility** among the ladies, even with varied years of experience. Later in life, I saw it as a shield for the insecurities within an environment where you constantly felt measured or checked against the males we worked with. Nobody had time to pull anyone else up when they were not sure of their own footing.
6. **Lack of mentorship and support networks.** There was a lack of female mentors or role models to mentor or guide one.

IT GENDER PARITY OBSTACLES

There have been notable strides since those years. Still, despite increasing efforts and ways to attract women, IT remains a male-dominated field, and women continue to face obstacles that make the journey toward a successful IT career difficult. These include the following:

Gender-Dividing Stereotypes

1. The societal stereotype that men are more suited to technical and logical tasks has limited female participation in these areas. Statements like "Girls are better at reading and writing, not technical subjects" and "That's a boy's toy,

girls don't play with building sets or robots" when strengthened by the purchase of dolls for a girl and Lego sets for a boy, are stereotypical statements and actions that reinforce gender divisions in academic and play choices. These discourage girls from exploring STEM-related toys and activities early on, limiting their interest in engineering or construction-related fields.

2. Workplace and career stereotypes: Here we have the saying that "Women are more suited to caring and nurturing roles like teaching or nursing." This stereotype pushes women toward traditionally "feminine" roles, limiting their participation in fields like engineering, finance, or technology.
3. Cultural Stereotypes: "Men should be the breadwinners, and women should stay home with the children." This reinforces gendered expectations of men and limits their involvement in caregiving, while also devaluing women's economic contributions. Another one is "Girls should focus on becoming good wives and mothers, not on careers." This reinforces the expectation that women should prioritize domestic roles over professional ambitions, limiting their opportunities for career advancement. It is worthwhile to note that these cultural stereotypes that deter women from considering careers in STEM are pervasive even in countries where there is full educational access.
4. Media and Cultural Stereotypes: Popular culture often depicts scientists, engineers, and tech innovators as men, reinforcing the stereotype that STEM is a "man's world." This lack of representation in media and film can discourage girls from imagining themselves in STEM roles. When women are shown in STEM roles, they are often portrayed in a way that emphasizes their gender rather than their abilities, reinforcing the notion that they are exceptions rather than the norm. "Men are the inventors and innovators; women support from behind." This creates a narrative where women are seen as secondary contributors to innovation, which devalues their potential in areas of invention and discovery and builds the perception that computers and technical skills are for males. Historian Maria Hicks and other researchers have highlighted how early advertisements for computers like the Commodore 64 and Apple II depicted boys using computers while women were either absent or relegated to passive roles, such as household duties or secretarial tasks.

These stereotypes reinforce harmful gender norms that limit the potential of women. In education, they can push girls away from STEM fields, while in the workplace, they create barriers to leadership and career growth for women. Tackling these stereotypes is essential for achieving gender equality in both personal and professional environments.

Other obstacles are listed as follows:

Lack of Role Models

1. Underrepresentation in STEM Professions: Women are underrepresented in many STEM professions, which means girls have fewer female role

models in these fields. Without visible examples of women excelling in STEM, young girls may be less likely to see themselves pursuing similar careers.
2. Limited Mentorship: The lack of female mentors in STEM fields can contribute to women feeling isolated or unsupported in these environments, making it harder for them to persist in STEM careers.

Confidence Gap and Perception of Abilities

1. Self-Doubt: Research has shown that girls, even in countries with gender parity in education, often underestimate their abilities in STEM subjects compared to boys. This lack of confidence can discourage girls from taking advanced STEM courses or pursuing STEM degrees.[3]
2. Perceived Complexity: STEM subjects are often seen as more challenging, and without encouragement or support, girls may be more likely to choose fields they perceive as less demanding, even if they are equally capable.

Bias in Classroom and Learning Environments

1. Teacher Bias: In some cases, teachers may (consciously or unconsciously) treat boys and girls differently in STEM classes. Boys might receive more encouragement or attention in math and science, reinforcing the stereotype that these subjects are more suited for them.
2. Curriculum and Materials: STEM education materials and examples may sometimes cater more to boys' interests or experiences, making it harder for girls to connect with the subject matter.

Workplace Discrimination and Career Progression

1. Hostile Work Environment: Women in STEM careers often face a "chilly" work environment, including subtle (and sometimes overt) discrimination. This may include bias in hiring, promotions, and pay, as well as lack of family-friendly policies, making it more difficult for women to thrive and advance in these fields.
2. Retention Issues: Even when women enter STEM professions, they are more likely to leave due to a combination of workplace discrimination, lack of work-life balance, and isolation in these environments.

Field-Specific Barriers

1. STEM Field Preferences: Even in countries where women have near-equal participation in higher education, they tend to be underrepresented in specific STEM fields like engineering, computer science, and physics. These fields are often perceived as highly technical and competitive, which can deter women from pursuing them.

2. Overrepresentation in Life Sciences: While women may pursue degrees in life sciences (such as biology or medicine) at rates comparable to or higher than men, they are less likely to enter engineering, technology, or physical sciences, where gender gaps remain wider.

Socialization and Early Influences

1. Toys and Activities: From a young age, girls are often steered toward toys and activities that promote creativity and caregiving (e.g., dolls, arts, and crafts) while boys are encouraged to play with toys that emphasize building, problem-solving, and technology (e.g., LEGOs, robots). This early socialization can shape girls' interests away from STEM subjects.
2. Parental Expectations: Parents may also have different expectations for their daughters and sons when it comes to STEM, consciously or unconsciously reinforcing gender norms that guide career aspirations.

Lack of Supportive Infrastructure

1. Institutional Barriers: Schools and universities may lack the support systems needed to help women succeed in STEM. This includes gender-sensitive teaching methods, mentorship programs, scholarships for women in STEM, and career development initiatives specifically tailored to address the unique challenges women face.
2. Inflexible Work Policies: STEM careers, particularly in tech and engineering, are often associated with long working hours and inflexible schedules, which may deter women, especially those balancing family responsibilities, from pursuing or staying in these fields.

Systemic Bias in Research and Academia

1. Gender Bias in Research Funding: Studies have shown that female researchers often receive less funding than their male counterparts, which can limit their ability to advance in STEM academia and build high-profile research careers.[4]
2. Gender Gaps in Publication and Recognition: Women are also less likely to be published in top-tier journals or recognized for their contributions to STEM fields, which may affect their career trajectories.

STRATEGIES TO OVERCOME BARRIERS

Addressing this issue requires systemic and societal changes, starting with early education.

Several changes to early education have been shown to increase the likelihood of girls pursuing IT or STEM careers. These changes aim to address gender stereotypes, increase engagement with STEM subjects, and create supportive environments that encourage girls to explore and excel in these fields.[5]

Here are some key changes:

Early Education

1. Curriculum Reform and Integration of STEM
   Schools that integrate STEM topics throughout the curriculum, rather than isolating them in specific subjects, expose girls to STEM concepts early on. Introducing engaging, real-world applications of STEM (such as coding, robotics, digital literacy programs, and experiments) can spark curiosity and interest in these subjects. Programs like *Computing at School in the UK and Robotics in the US are such examples.*
2. Gender-Neutral Learning Materials
   Traditional learning materials often reinforce gender stereotypes, portraying boys as more suited for science and technology while girls are shown in less technical roles. Updated, gender-neutral learning materials present both genders as equally capable in STEM subjects, helping girls see themselves in these roles.
3. Female Role Models and Mentorship
   Introducing female role models in STEM fields during early education has a strong impact on girls' aspirations. Seeing women who have succeeded in STEM careers helps girls believe that they too can excel in these fields. Programs like *STEM Ambassadors* in the UK allow female scientists, engineers, and technologists to visit schools, share their experiences, and inspire girls to pursue similar paths.
4. Hands-on, Inquiry-Based Learning
   Girls often thrive in environments where they can explore and experiment with concepts rather than passively receive information. Hands-on, inquiry-based learning that encourages curiosity, problem-solving, and collaboration allows girls to actively engage with STEM subjects in a way that feels more relevant and accessible. Programs that focus on building things (robotics, science experiments, coding challenges) can give girls confidence and interest in these fields.
5. Addressing Implicit Bias in Teaching
   Teachers play a crucial role in shaping how girls perceive their abilities in STEM. Providing teachers with training to recognize and counteract implicit gender bias in the classroom can have a significant impact and help educators avoid reinforcing stereotypes while ensuring that all students have equal opportunities to succeed in STEM subjects.
6. STEM-Related Play in Early Childhood
   Introducing STEM concepts through play at a young age—such as with building blocks, logic puzzles, or simple science kits—helps girls develop problem-solving skills and spatial awareness, both of which are important in STEM. One can have the proverbial dolls for girls, but getting educational toys like chemistry sets leads to exposure to realms outside the norm. These days, doll sets have cars or planes where doors can fall off hinges and need to be fixed. The repair does not necessarily have to be a male activity.

7. Supportive Classroom Culture
   Creating a classroom culture that fosters collaboration and values different ways of thinking can make STEM subjects more appealing to girls. This environment should encourage questions, experimentation, and failure as a learning process, which reduces anxiety about not "getting it right" immediately. Programs that focus on teamwork, such as group projects in science or coding competitions, allow girls to contribute in meaningful ways without the fear of being singled out.
8. Single-Sex STEM Programs
   Research has shown that 93% of students in all-girls schools felt encouraged to develop confidence in math and science, compared to 48% in coeducational settings. Single-sex STEM programs or classes can encourage girls to explore these subjects without the pressure of competing with boys or adhering to gender expectations.[6] STEM subjects. All-girls coding camps, such as *Girls Who Code*, create safe spaces for girls to learn and excel in programming and technology, free from the influence of stereotypes.
9. Early Exposure to Engineering and Technology
   Girls are often not introduced to engineering and technology until later in their education, at which point they may have already internalized the belief that these fields are not for them. By introducing these subjects early, through creative, practical activities, schools can show girls that they are equally capable in these areas. Companies like "Lego" took this challenge on in 2021, stating that they will work harder to remove gender biases from their products and marketing, citing the results of a worldwide survey in a study, "Ready for Girls" conducted by the LEGO Group in partnership with the Geena Davis Institute on Gender in Media, which found that general attitudes toward kids' play and creative careers remain "unequal and restrictive." This led them to create gender-friendly themes and pieces for their sets.
10. Parental Involvement
    Encouraging parents to engage their daughters in STEM activities at home can significantly influence girls' attitudes toward these fields. Parents can help break down gender stereotypes by providing STEM toys, encouraging exploration, and talking positively about the flexibility and portability of STEM careers for women in a way I would call "Demystifying or normalizing the sciences."

Mentorship and Support

1. Direct mentorship
   This has been one of the most impactful strategies for overcoming the challenges I faced as a woman in STEM. Later in my career, I was fortunate to have male and female mentors who provided guidance, shared their experiences, and offered support. Mentorship does not only provide professional advice; it is also a source of confidence and empowerment, enabling

women to navigate corporate cultures that may be unfamiliar or unwelcoming. Mentorship is essential, not only for providing guidance but also for increasing visibility for women in IT leadership.

2. The Role of Women in Shaping Future STEM Leaders
   The responsibility of shaping the future of women in STEM does not only fall on external forces, such as policy changes or corporate initiatives. Women in STEM, especially those in leadership positions, have a crucial role to play. Through mentoring, advocacy, and setting positive examples, they can open doors for future generations by encouraging younger generations to believe that they too can break into the field and succeed. As we look to the future, it is essential that women in STEM fields work to influence organizational culture. Creating inclusive and diverse workplaces not only benefits women but also enhances innovation and creativity, leading to better business outcomes.[7]
3. STEM Support Networks
   The creation of a more inclusive environment in STEM requires intentional efforts to mentor and sponsor women at different stages of their careers. Initiatives like establishing mentorship programs, organizing workshops for career development, and offering flexible work arrangements can help create an environment that supports women as they navigate STEM careers. The creation of these networks emphasizes the importance of peer mentorship and highlights how vital it is for women to create opportunities for mutual support, providing guidance, perspective, and career insights. A report by McKinsey & Company emphasized that women with mentors and sponsors were more likely to be promoted and less likely to leave their companies.[8]

Continuous Self-Improvement and Self-Development

I believe that confidence and competence go hand-in-hand. Women must build their technical and leadership skills while simultaneously working on self-confidence. In technology, advancements happen at breakneck speed, and staying relevant requires a commitment to learning. Certifications, training, and hands-on experience have all been part of my journey toward ensuring I remain on the cutting edge of IT and, more recently, risk management.

Women in STEM must continually invest in their education and skills. Whether through formal education or on-the-job learning, this process helps build confidence and ensures that women are prepared to lead in their respective fields.

Challenging Stereotypes

Challenging stereotypes requires a firm belief in one's capabilities and a willingness to assert those capabilities even in the face of doubt. In the IT sector, where technical expertise is paramount, I have found it essential to demonstrate my competence repeatedly. By consistently delivering high-quality work and communicating effectively, I have been able to change the perceptions of those who might have initially questioned my expertise in the IT space.

Networking

The key to successful networking for women in IT is to be proactive, engage in a variety of platforms, and seek out opportunities for both formal and informal connections. By joining women-focused tech communities, attending events, leveraging online platforms, and engaging in mentorship or thought leadership, women in IT can build strong networks that support career growth, leadership opportunities, and skill development.

1. Join Women in IT Networks and Communities.
2. Attend Conferences and Networking Events.
3. Utilize Online Platforms and Social Media.
4. Join Corporate Women's Networks.

Tenacity

Persistence is crucial. Success in any career, especially one in STEM, requires the tenacity to take on and push through challenges, whether they are external obstacles or internal doubts or sometimes, even ability. Women must remain steadfast in their pursuit of career advancement, if that is what they want, and not allow temporary setbacks to derail their long-term goals.

Government Initiatives

In the early 1990s, there was a push in many parts of the UK to increase the number of women in computing, driven by both national and local initiatives to address the underrepresentation of women in technology fields. During that time, efforts were being made to encourage greater diversity and inclusion in STEM. Several measures were implemented across the UK, including Sunderland, to increase the number of women in computing. This was a program I benefited from, and this push included:

1. National curriculum reforms.
2. Equal Opportunities Policies: a greater national emphasis on gender equality in the workplace.
3. National campaigns like the Women Into Science and Engineering (WISE) Campaign aimed to encourage females to pursue STEM careers.
4. Local university efforts.
5. Community and Outreach Programs: Workshops and IT training for women aimed to provide practical computing skills, often targeting those returning to the workforce or considering a career change.

6. Employment and training schemes
7. Girl-Focused Computing Clubs: Some schools and youth organizations have started to create computing clubs aimed specifically at girls, designed to give young women hands-on experience with computers and programing in a supportive, non-competitive environment.
8. Women in Technology Role Models: This included inviting female guest speakers to schools and universities to share their experiences in computing careers.
9. Mentorship Programs: Mentorship schemes where women already working in IT or studying computing are paired with younger girls or students to provide guidance and encouragement.
10. Local Business Engagement: Businesses were encouraged to support women entering the computing workforce by offering internships, job placements, and flexible working conditions.

These foundational efforts laid the groundwork for continued initiatives to promote women in technology in the decades to follow.

DATA EXCERPTS

Following is pertinent data that we have seen on female representation in STEM education and in STEM employment.

STEM Education

1. Enrollment Gaps: According to the 2023 report, women continue to be significantly underrepresented in STEM fields at both the secondary and tertiary education levels. Only 30% of students enrolled in STEM-related programs globally are women.
2. Fields of Study: While women are gaining ground in life sciences and health, they remain underrepresented in fields like computer science, engineering, and information technology.
3. Barriers to Entry: Social stereotypes, gender norms, and lack of female role models were identified as key reasons for this imbalance. These factors often deter girls from pursuing STEM from an early age.

STEM Employment

1. Workforce Representation: Despite a growing presence in the workforce, women make up only 28% of the global STEM workforce (as of 2023). This gap is particularly notable in high-tech industries like artificial intelligence, cloud computing, and cybersecurity.
2. Leadership Positions: In 2023, only 16% of leadership roles in AI and related fields were held by women, further reflecting the lack of female representation in advanced STEM careers.

3. Wage Gaps: Women in STEM fields continue to face wage disparities. On average, they earn less than their male peers, with gender gaps in earnings as high as 20% in some regions.
4. Regional Variations: Countries with higher gender parity, such as Nordic nations, have made better progress in STEM inclusion, whereas developing economies show larger disparities, especially in STEM employment opportunities for women.

According to the 2023 Global Gender Gap Report, Iceland ranked highest in closing the gender gap, while the STEM-related gender gaps remain significant even in top-ranked countries. The report stresses that at the current rate of progress, it will take an estimated 131 years to close the global gender gap across all sectors, including STEM.

Initiatives and Recommendations

The report emphasizes the need for policy interventions, mentorship programs, and flexible work environments to encourage female participation in STEM fields. Governments, educational institutions, and the private sector are called to invest in initiatives that promote gender equality in STEM education and careers.

This gender disparity in STEM is an ongoing global issue, and despite gradual improvements, significant gaps remain in both educational enrollment and professional representation.

Addressing these issues requires concerted efforts, including changing cultural norms, creating more inclusive educational environments, providing strong mentorship, and ensuring supportive work policies that encourage women to enter and thrive in STEM fields, especially as data provides quite dismal findings even in countries where there are inclusive advances:

1. Women comprise less than 25% of workers in cloud computing and engineering jobs.
2. Thirty nine percent of all jobs were held by women, as reported on the LinkedIn platform. However, women represent only 12% and 15% of the workers in cloud computing and engineering occupations, respectively.
3. The United States has slightly higher proportions of women in cloud computing (14%) and engineering (23%) jobs, which mirrors the overall higher percentage of all (47%) jobs held by women. Still, Canada (16%), Australia (19%), and India (20%) also stand out for having slightly higher female representation in engineering occupations.

FINDINGS/CONCLUSION

Breaking STEM gender barriers requires a multifaceted approach. Women starting their careers in IT must focus on seeking out mentors who can support them and advocate for their careers. They must engage in continuous self-improvement and network strategically. Women who reach positions of authority must use their influence to drive inclusive policies and hiring decisions and work to create psychologically safe environments where the next generation of female leaders can thrive.[9]

The IT field offers incredible opportunities for women who can navigate its challenges. The opportunities are numerous and, by building a strong support system, acquiring relevant skills, and staying resilient in the face of adversity, women can grow in the industry and lead it toward a more inclusive future and resilient workforce that can meet the complex demands of the digital age. Ultimately, a more gender-balanced IT industry benefits businesses and society, driving greater progress and opportunities for all.

DISCUSSION

Barriers to females accessing education, poor school retention, lack of opportunities for higher education, cultural norms, economic factors, and early marriage may hinder the education of females and therefore STEM education as well. One would think that with access to education, the percentages would be better. However, even in countries where there is near-total gender parity in educational access, there is still a STEM gender disparity gap, so the reasons are endemic whether education is available or not.

Increase in numbers and the success of women in STEM is achievable but requires systemic change at individual, organizational, and government levels as described in the prior sections.

SOCIAL/PRACTICAL IMPLICATIONS

Knowing some of the social and practical implications of gender disparity can further drive efforts to reduce it.

1. **Social Implication**
   a. Women in STEM can inspire and uplift communities and promote social equity, particularly in regions where girls' education or career prospects are limited.
   b. Women in STEM are powerful examples of how education and opportunity can drive social mobility, transform communities, and provide equal access to high-growth and high-reward opportunities for scientific progress, economic development, and global well-being.
2. **Practical Implication**
   a. When women succeed in STEM, they often give back to their communities by starting initiatives that promote STEM education for girls, offering mentorship, and advocating for policy changes that support women and minorities in science and technology fields. This leads to a cycle where more girls are inspired to follow similar paths.

b. Women in STEM promote the development of more inclusive educational and workplace policies, as governments and organizations may be encouraged to implement scholarship programs and flexible work environments.

REFERENCES

1. World Economic Forum. (2023). Global gender gap report 2023.
2. Cech, E. A., & Blair-Loy, M. (2019). The changing career trajectories of women in STEM. *Sociology of Education*, *92*(1), 58–78.
3. OECD. (2018). *Global gender disparities in STEM: Self-confidence and aspirations.* OECD.
4. Tamblyn, R., McKenzie, K., & Seltzer, M. (2020). Gender and the funding of research: A systematic review. *Academic Medicine*, 95(12), 1932–1938.
5. Wang, M.-T., & Degol, J. L. (2019). Improving gender equity in STEM through early educational interventions: A review of the literature. *Educational Psychology Review*, *31*(4), 773–801.
6. Cornell University. (2016). *Gender and confidence in mathematics and science: The role of single-sex schooling.*
7. Dasgupta, N., & Stout, J. G. (2014). Girls and women in science, technology, engineering, and mathematics: STEMing the tide and broadening participation in STEM careers. *Policy Insights from the Behavioral and Brain Sciences*, *1*(1), 21–29.
8. McKinsey & Company. (2020). *Women in the workplace 2020.*
9. Murphy, M. C., Steele, C. M., & Gross, J. J. (2007). Signaling threat: How situational cues affect women in math, science, and engineering settings. *Psychological Science*, *18*(10), 879–885.

Index